# Ceramic Materials - Present and Future

*Edited by Amparo Borrell*
*and Rut Benavente*

Published in London, United Kingdom

Ceramic Materials - Present and Future
http://dx.doi.org/10.5772/intechopen.1000320
Edited by Amparo Borrell and Rut Benavente

Contributors
Marina Magro Togashi, Claudia P. Fernandez Perdomo, Guido Link, Jhon Jelonnek, Ruth Herta G.A. Kiminami, Eugeni Cañas, Rut Benavente, Amparo Borrell, Ma Dolores Salvador, Amparo Borrell, Rut Benavente, René M. Guillén, María D. Salvador, Vaclav Pouchly, Martina Ilcikova, Richard Krumpolec, Rodrigo Moreno, Cecilia Granados-Miralles, Matilde Saura-Múzquiz, Henrik L. Andersen, Cécile Autret-Lambert, Samir Merad, Sonia De Almeida-Didry, Olivier Motret, François Gervais, Angelina Stoyanova-Ivanova, Stanislav Slavov, Bin Lin, Jingguo Zhou, Jinhua Wei, Haoji Wang, Tianyi Sui, Leane Priscilla Bonfim Sales, Aline Figueiredo da Nóbrega, Iranilza Costa da Silva, Ana Cecília Vieira da Nóbrega, Arnaldo Manoel Pereira Carneiro, Fabiola Luana Maia Rocha, Diego de Paiva Bezerra, Seung Jin Kim, Hyun Ah Kim



First published in London, United Kingdom, 2023 by IntechOpen
IntechOpen is the global imprint of INTECHOPEN LIMITED, registered in England and Wales, registration number: 11086078, 5 Princes Gate Court, London, SW7 2QJ, United Kingdom

British Library Cataloguing-in-Publication Data
A catalogue record for this book is available from the British Library

Additional hard and PDF copies can be obtained from orders@intechopen.com

Ceramic Materials - Present and Future
Edited by Amparo Borrell and Rut Benavente
p. cm.
Print ISBN 978-1-83769-703-8
Online ISBN 978-1-83769-702-1
eBook (PDF) ISBN 978-1-83769-704-5

# Meet the editors

Dr. Amparo Borrell has a degree in physics and a Ph.D. in Materials Science from the Spanish National Research Council (CSIC). From 2018 to 2022 she was a Ramón y Cajal researcher. At present, she is a Senior Lecturer of Materials Science and Engineering at the Polytechnic University of Valencia, Spain. Her main research lines are ceramics processing by non-conventional sintering techniques. She is the author and co-author of about ninety-five papers in international scientific journals and has participated in twenty-six research projects. She has participated in 130 congresses and several technology transfer activities that have generated three patents. She has been a visiting researcher at the University of São Paulo, Brazil, and the University of Modena e Reggio Emilia, Italy.

Dr. Rut Benavente is a senior lecturer at the Polytechnic University of Valencia, Spain. She has a degree in chemistry from the University of Valladolid, Spain, and a Ph.D. in Materials Science from the Polytechnic University of Valencia. Her main scientific activity is focused on the science and engineering of advanced ceramic materials with specific functionalities, covering different aspects ranging from the processing of materials to the study of mechanical and microstructural properties. She is the author of thirty scientific publications and has collaborated in more than forty contributions to national and international scientific congresses.

# Contents

# Preface

In a constantly evolving world, the quest for advanced materials that can revolutionize technology, industry, and our everyday lives has become a crucial objective for the scientific and technological community. Ceramic materials, with their unique combination of properties, have emerged as promising candidates to play a fundamental role in the future of various applications. This book is a multidisciplinary exploration of recent research in the field of ceramic materials, from their incorporation in lime mortars to their application in eco-friendly coatings and high-permittivity core-shell nanostructures.

One of the most intriguing developments in ceramic materials engineering is the use of lime mortars containing ceramic materials as pozzolans. These mixtures are intended to strengthen the properties of construction materials, increasing their durability and sustainability. By combining tradition with innovation, these materials could be key to the construction of the future.

In the realm of technology, permanent magnets based on hard ferrite ceramics have proven to be essential in a variety of applications, from electric motors to medical devices. The book explores the latest advances in this area and how these magnets are driving the efficiency and effectiveness of a wide range of technologies. The search for materials with exceptional dielectric properties for use in capacitors has led to the development of core-shell nanostructures in colossal permittivity materials. The book also investigates the synthesis of these nanostructures, as well as their dielectric properties, and how they could revolutionize the electronics industry. Furthermore, the impact of ceramic materials on our everyday lives cannot be overlooked. As such, the book delves into how the incorporation of ceramics into fabrics is yielding textiles with unique thermal characteristics, providing a warm-cool feeling. This breakthrough may have surprising applications in protective clothing and daily comfort.

Sustainability is a hot topic today, and the production and application of eco-friendly ceramic coating materials are key research areas. The book explores how ceramic materials can contribute to environmental protection while simultaneously enhancing the durability and performance of various products and structures. In industry, process optimization is essential for efficiency and product quality. The book analyzes how research into grinding force and process optimization in woven ceramic matrix composites opens new possibilities in the manufacturing and performance of advanced materials.

Finally, the book discusses the densification kinetics and in situ electrical resistivity measurements of hematite nanopowders during high-frequency microwave sintering. This study sheds light on how ceramic materials can be processed more efficiently with improved electrical properties.

This preface is just a glimpse into the exciting variety of ongoing research in ceramic materials. We stand on the threshold of discoveries that can change how we build, technologically advance, and live. Ceramic materials are destined to play a central role in this exciting journey into the future, and this research is a testament to their continued importance in modern science and technology.

**Amparo Borrell and Rut Benavente**
Polytechnic University of Valencia,
Institute of Materials Technology,
Valencia, Spain

Section 1

# Advanced Ceramics Obtained by Rapid Processing

Chapter 1

# Densification Kinetics and *In Situ* Electrical Resistivity Measurements of Hematite Nanopowders during High-Frequency Microwave Sintering

*Marina Magro Togashi, Claudia P. Fernandez Perdomo, Guido Link, Jhon Jelonnek and Ruth Herta G.A. Kiminami*

## Abstract

The aim of this work was to study the densification kinetics and to evaluate the electrical resistivity of hematite nanopowders (30 nm) during sintering by high-frequency microwave dilatometry (30 GHz) in multimodal cavity, as the purpose of providing new reference data on the kinetic behavior of the densification of high dielectric loss ceramic materials when subjected to microwave sintering, which are few in the literature to date. To analyze the densification kinetics, non-isothermal methods were used, such as the classic models of Woolfrey-Bannister and Dorn for the initial stage of sintering, and the model of Wang-Raj for the intermediate stage, both at heating rates ranging from 5 to 20°C/min. The results show an extreme very low activation energies, both for the initial stage (39–66 kJ/mol) and for the intermediate stage (68 kJ/mol), proved to be sufficient for material densification with less grain growth, compared to lower frequencies (2.45 GHz). In addition, the *in situ* resistance measurements revealed a decrease in electrical resistivity as a function of material densification, as well as a semiconductor behavior of the sintered hematite.

**Keywords:** microwave sintering, high frequency (30 GHz), densification kinetics, initial and intermediate stage, activation energy

## 1. Introduction

The need for a better understanding of the sintering kinetics of microwave-assisted sintering of ceramics is growing. It is an unconventional process that has shown to be very efficient in the production of advanced high-density ceramics. It leads to a significant reduction in sintering time, as a result of enhanced diffusion processes based on non-thermal effects [1]. Studies on the sintering kinetics of the microwave-assisted process of ceramics sintering, particularly if considering hematite materials, are still scarce in the literature. The few studies reported in the literature adopt the classic sintering models used for conventional sintering. This is valid for both the isothermal sintering, which assesses the grain growth kinetics, and the non-isothermal sintering, which assesses the densification kinetics. The studies are done mainly at a microwave frequency of 2.45 GHz and in materials with an average particle size at a micrometric scale [2–5]. Thus, the viability of using these classic models has encouraged the use of dilatometers heated by microwave radiation, both at 2.45 and at 30 GHz. This allows to evaluate the instantaneous linear shrinkage and enable studies of the kinetics behavior during sintering at different frequencies [2, 4, 6, 7]. Gyrotrons are used as the high-power microwave source at 30 GHz. The operation at 30 GHz is equivalent to a wavelength in free space of ~10 mm. It offers a huge advantage for sintering such as increased power absorption. It enables the sintering of materials with low dielectric losses, which eliminates the need of using susceptors as the absorbed power density ($P_{abs}$) absorbed by the ceramics is directly proportional to the frequency [8, 9]. Eq. (1) shows the dependence of the absorbed power density from the operation frequency of the microwave [10].

$$P_{abs} = \sigma E^2 = \omega\varepsilon' \tan\delta E^2 = \omega\varepsilon'' E^2 \qquad (1)$$

In (1) $\sigma$ is the effective material conductivity, $E$ is the electrical field, $\omega$ is the angular frequency of the wave. $\varepsilon'$ and $\varepsilon''$ correspond to real and imaginary part of effective permittivity, respectively, and $\tan\delta$ is the loss tangent of the material.

In the literature, some works report on using microwaves operating at 30 GHz to assist ceramic sintering such as the thin film of barium and strontium titanate reported by Paul et al. [11]. The authors observed that sintering is fast with a reduction of the sintering time of approximately 67% if compared to conventional sintering. Moreover, the authors observed a lower incidence of defects, better homogeneity, and quality of films sintered by microwave. Birnboim et al. [12] made a comparative study of ZnO sintering at three different frequencies (2.45, 30, and 83 GHz) without any type of auxiliary heating or susceptor material. The authors monitored the interior and surface temperatures of the samples. It was observed that at 2.45 GHz, the interior temperature was higher than on the surface due to the heat losses from the surface, while at 30 and 83 GHz, the exterior temperature was higher, especially at high heating rates. The authors associated this to a decrease of the penetration depth when frequency is increased. These thermal gradient differences between the interior and surface depend on factors such as thermal and dielectric properties, the frequency, and heating rate. In 2013, Sudiana et al. [13] studied the alumina densification in microwave furnaces at 30 GHz and at 300 GHz. A raise in the densification was seen when using microwave-assisted sintering at 30 GHz if compared to the conventional sintering. Besides, the authors observed that the diffusion rate at 30 GHz was higher than at 300 GHz in all studied temperatures, and the grain sizes were similar in both processes at corresponding densities.

Dilatometer measurements of ceramics sintering provide important information related to the curve of linear thermal shrinkage in order to determine the temperature of the onset of shrinkage, ranges of initial and intermediate stages temperatures, and maximum linear shrinkage temperature. Link et al. [8] studied the sintering kinetics of nanometric zirconia stabilized with yttria in a at 30 GHz gyrotron system including a dilatometer. The results reveal a lower temperature for the onset of shrinkage as compared to the conventional heating as well as the retention of the nanoscale microstructure at the end of the process. Furthermore, the authors studied lead zirconate titanate (PZT) and observed lower sintering temperatures and a lower PbO loss by evaporation which allows for a better piezoelectric performance of the material. Thumm et al. [9] observed an advantage in processing of advanced materials such as submicron YSZ, PZT, dental ceramics, and $SiO_2$ using microwave-assisted sintering at 30 GHz.

According to the data provided in the literature, if comparing the activation energy for densification during the microwave-assisted with conventional sintering, normally a reduction of the activation energy is observed during the heating of the material with microwaves. This reduction is normally associated with the non-thermal effects of the electromagnetic radiation. The hypothesis with better acceptance in the scientific community was postulated by Rybakov et al. [10, 14, 15]. An additional driving force, the so called ponderomotive driving force, acts in the mass transport of the ceramic material once a nonuniform electromagnetic field generates oscillatory movements of the charged particles. This additional force is of a nonthermal nature, resulting in a potential reduction of activation energy during sintering.

Link et al. [16] studied metallic compacts in the same dilatometry system for *in situ* electrical resistivity measurements as used in this report. These results allow the association to the electrical resistivity behavior as a function of the microwave sintering process. It can also be potentially used for ceramic materials. Pomar et al. [17] studied the microtubes of iron oxide synthesis by thermal oxidation of the hematite process. The process was monitored by *in situ* electrical resistivity measurements. The authors observed that during heating the values of electrical resistivity increase, as expected. Besides, they observed that during the formation of the microtubes, in the thermal oxidation process around 500°C, there was an abrupt change in these values, increasing by two orders of magnitude.

Hematite material is an important n-type semiconductor with a band gap in the range of 1.9 to 2.2 eV [18]. It is a material of great interest due to its properties that enable applications in several fields such as photochemical cells, catalysis, and gas sensors [19]. The hematite has relatively high dielectric losses. Consequently, it is a good absorber for the electromagnetic radiation. As reported by Ramya et al. [20], the loss tangent of the hematite measured with different initial particle sizes in the range of 19 to 33 nm was 0.704–1.463, respectively, in trials performed at 40°C and 1 kHz. Data of dielectric values of hematite and measurements in other frequencies are still scarce.

Other studies were performed on the hematite under microwave heating, such as reported by Togashi et al. [7]. The densification kinetics of the hematite nanopowders was evaluated using the linear thermal shrinkage curves in the analysis, obtained in a microwave-heated dilatometer at 2.45 GHz aided with a microwave susceptor and by dilatometry with conventional heating. The main conclusion of the study was the reduction of the activation energy for the initial stage and the intermediate stage of the microwave sintering. In the initial stage, the reduction is in the order of 10% and for the intermediate stage of 56.3%. This provides a better sintering kinetics and a reduction of the grain growth if compared to conventional heating. These results showed success in evaluating the sintering kinetics in samples sintered by microwave at 2.45 GHz, enabling a better thermodynamic understanding during sintering of the hematite at a low frequency (frequency 10× lower than in this study).

Thus, this work aims for the study of the sintering kinetics and the *in situ* electrical resistivity measurements of hematite nanopowders sintered by microwaves in the millimeter wave range.

## 2. Materials and methods

Hematite nanopowders ($Fe_2O_3$) with an average particle size of 30 ± 2 nm were synthesized by a modified sol-gel method, as described by Togashi et al. [7]. For the dielectric characterization, the dielectric permittivity was measured with a cavity perturbation system, as the similar to the experimental scheme of Soldatov et al. [21]. The frequency was 2.45 GHz, and the temperature was varied from room temperature to 730°C. The compacts were in cylindrical shape with 7.8 mm of diameter and 10 mm of width. The penetration depth ($D_p$) was calculated by Eq. (2) [22].

$$D_p = \frac{1}{2\left(2\pi f\sqrt{0,5\mu_0\varepsilon_0\varepsilon_r'\left(\sqrt{\left(1+\left(\frac{\varepsilon_r''}{\varepsilon_r'}\right)^2\right)}-1\right)}\right)} \tag{2}$$

$f$ is the frequency of microwave operation, $\mu_0$ is the magnetic permeability and $\varepsilon_0$, the permittivity, both in vacuum. $\varepsilon_r'$ is the relative real permittivity and $\varepsilon_r''$, the relative imaginary permittivity, measured in this case with the perturbation cavity system.

For the dilatometric experiments, a microwave field at 30 GHz is used. Hematite powder compacts in a rectangular prismatic shape, with approximately dimensions of 10×3×3 mm, were pressed in a cold uniaxial press (P O Weber – PW40 model) with 330 MPa of pressure. It assures similar green densities of approximately 40% of theoretical density. The full density of hematite is 5.27 g/cm$^3$ according to JCPDS-ICCD card number 89-0596. For the *in situ* electrical resistivity, the nanopowder was pressed in cylindrical shape with 5.5 mm in diameter and the length was 3 mm. Again, the density was 40% of theoretical density in all samples. Both experiments were performed in triplicate.

Both dilatometric and *in situ* electrical resistance measurement tests assisted by microwaves at 30 GHz were performed using a commercial dilatometer (Linseis GmbH, Germany) with alumina rod and support. It was coupled with a microwave furnace in a gyrotron system with maximum power of 15 kW, as described by Link et al. [15, 23]. **Figure 1(a)** schematizes the system. The cavity has a hexagonal shape and a rotational agitator on top to allow a more homogeneous field distribution. The heating rates used in this research were 5, 10, 15, and 20°C/min, up to 1200°C and a dwell time of 1 minute. The temperature was measured by a type-S thermocouple with a direct thermal contact to the sample (**Figure 1(b)**). **Figure 1(c)** schematizes the *in situ* four-wire electrical resistance measurements.

The *in situ* electrical resistance trials were performed with a heating rate of 20°C/min, and the measurements were performed using the four-wire multimeter method (Keithley 2002). To do so, two more electrodes were placed, each at one end of the sample, making a "sandwich" with the sample, as outlined in **Figure 1(c)**. All the experiments were carried out in air atmosphere.

For the first cycle, using the green sample, a heating rate of 20°C/min and a controlled cooling rate of 13°C/min were applied. Same rates were used for the second cycle of *in situ* electrical resistance measurements, where the samples are already sintered. The electrical resistivity (ρ) was calculated by Eq. (3):

$$\rho = \frac{Rr^2\pi}{l} \tag{3}$$

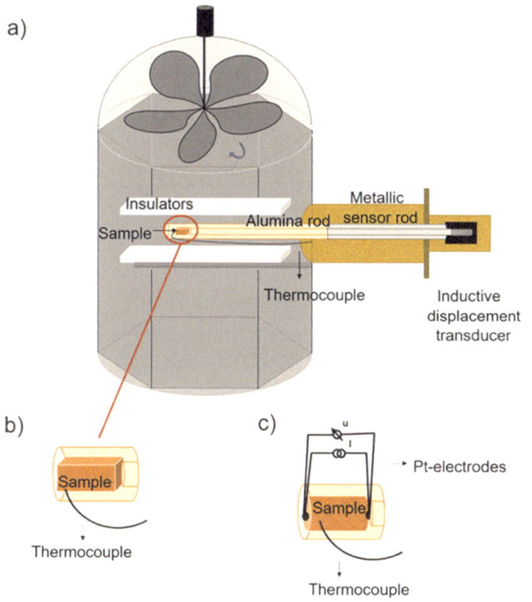


**Figure 1.**
*Scheme of the dilatometry system coupled with a microwave furnace in a gyrotron system (adapted figure from Link et al. [23]) (a), representation of the temperature measurements for the dilatometric tests, using the S-type thermocouple (b) and for the* in situ *four-wire electrical resistance measurement (c), in which, in addition to the S-type thermocouple, two more platinum electrodes were used.*

$R$ is the electrical resistance, in this study measured *in situ*, $r$ is the radius of the cylindrical sample and, $l$ its length, both for the green ceramic compact. In this case, $r$ was approximately 2.8 mm and the length, 2 mm.

The analysis of the sintering kinetics of the hematite nanopowders was performed using Woolfrey-Bannister and Dorn nonisothermal models for the initial stage of sintering [24] and Wang-Raj constant heating rate model, for the intermediate stage of sintering [25]. The calculation of the activation energy for densification in each of these stages of sintering was carried out and compared with the data in the literature based on conventional heating and microwave hybrid heating at 2.45 GHz.

All the samples were carefully cut in transversal direction into two separate parts after the dilatometric tests at the different heating rates (5, 10, 15, and 20°C/min). The cross section was grinded and polished. The 5 and 20°C/min heating rate samples were analyzed using an optical microscope (Olympus – BX41M LED) to evaluate the microstructural homogeneity and potential differential shrinkage of the sample. And a thermal etching at a temperature 50°C below the sintering temperature, using the same heating rate of the dilatometry, was made for all samples, in order to evaluate the grain size, using a high-resolution scanning electronic microscope (HR-SEM microscope Hitachi, S-4800 type). The average grain size was estimated with the ImageJ software. The measurements of apparent density were performed in ethanol adopting the immersion method, using the principle of Archimedes.

## 3. Results and discussion

The dielectric characterization measurements at 2.45 GHz of the hematite with 30 nm of initial particle size are presented in **Figure 2(a)**. The heating cycle was from

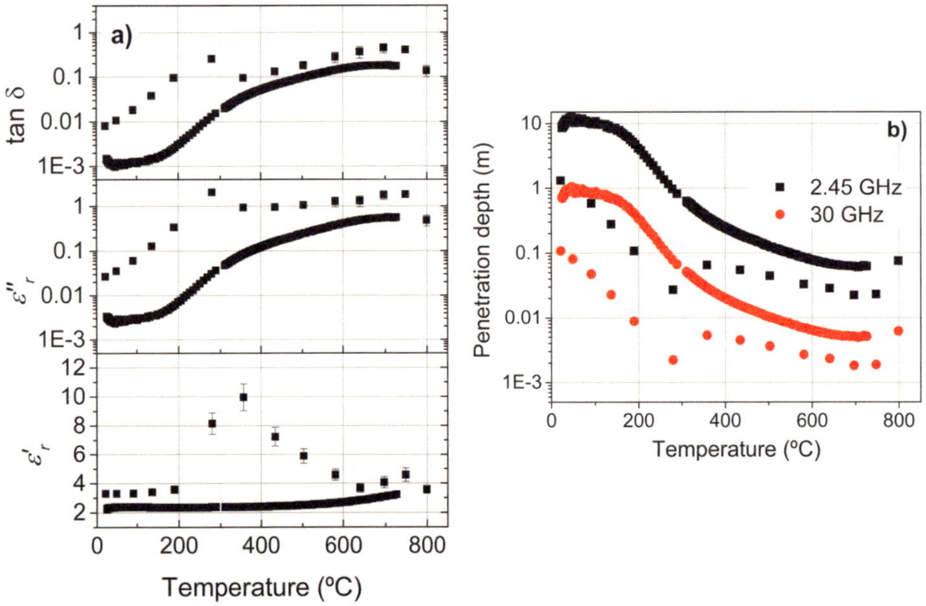


**Figure 2.**
*Dielectric characterization of nano-hematite by the perturbation cavity system at 2.45 GHz (a), measured penetration depth for 2.45 GHz and extrapolation curve for 30 GHz (b).*

room temperature to 730°C. In **Figure 2(b)**, it is presented the penetration depth calculated for 2.45 GHz and an extrapolation to 30 GHz. This extrapolation was made, considering that the permittivity values and loss tangent do not change with the higher frequency. The penetration depth is the distance of the material surface to the distance in which the field magnitude decreases in a factor of 1/e.

By the results, it is possible to observe that the permittivity, especially in the imaginary contribution, increases with the increase in temperature. The penetration depth decreases with the increase in temperature. In this case, at approximately 350° C, the hematite starts to absorb better the radiation at 2.45 GHz, characterizing the critical temperature of the material. Up to 680°C, there is an asymptotic behavior of the tan δ curve. The extrapolation curve shows the behavior of the hematite when sintered at high frequency. The penetration depth values would decrease, and it is expected that the absorbed power by the material increases. This results in better interaction of the hematite with the electromagnetic field with the increase in operating frequency.

### 3.1 Microwave dilatometry at 30 GHz

The microwave dilatometry results achieved with high frequency (30 GHz) sintering of hematite nanopowders with an average particle size of 30 ± 2 nm at different heating rates: 5, 10, 15, and 20°C/min, a final temperature of 1200°C, and dwell time of 1 minute are shown in **Figure 3(a)**. **Figure 3(b)** shows the zoom of the temperature of the onset of shrinkage, highlighting that the region $\Delta l/l_o$ becomes negative.

**Figure 3(b)** reveals the onset of shrinkage at around 200°C, when the Δl/lo values are negative. The temperature of maximum linear shrinkage rate corresponds to the

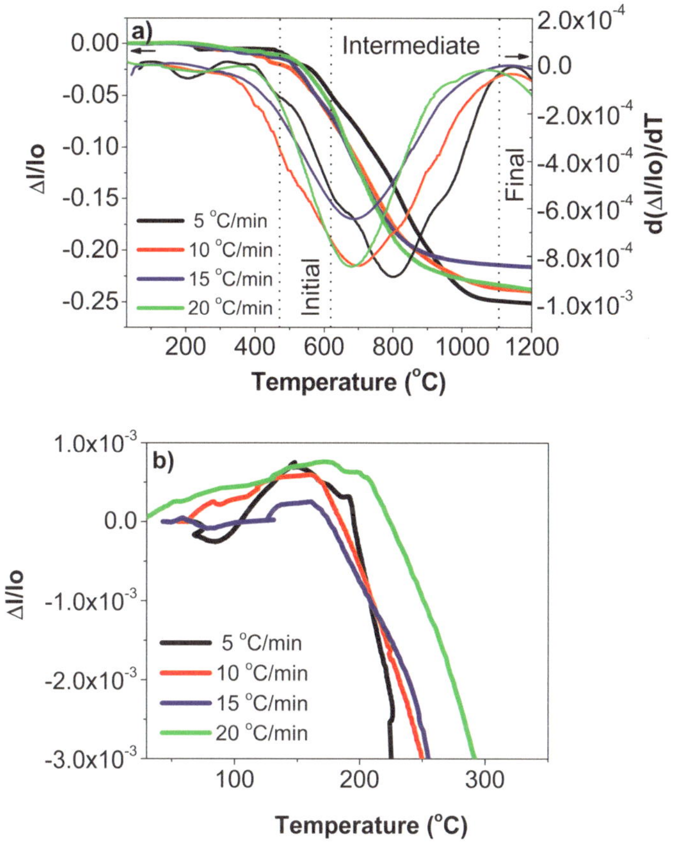


**Figure 3.**
*Microwave-assisted dilatometry in high-frequency (30 GHz) curves of the nanometric hematite powder (a) and temperature onset of shrinkage (b) at the different heating rates.*

minimum in the curves of the first derivative, which according to **Figure 3(a)**, is found in the range of 700–824°C. There is the tendency of lower values of maximum shrinkage rate temperatures with increasing heating rate, as presented in **Table 1**. In **Table 1**, the values of the temperature of onset of shrinkage, range of temperature of the initial stage, maximum linear shrinkage rate temperature, range of intermediate stage of sintering, and total shrinkage in the different heating rates are presented.

| Heating rate (°C/min) | Temperature of onset of shrinkage (°C) | Initial stage of sintering (°C) | Maximum linear shrinkage rate temperature (°C) | Intermediate stage of sintering (°C) | Total shrinkage (%) |
|---|---|---|---|---|---|
| 5 | 199 ± 8 | 521–626 | 824 ± 28 | 626–1100 | 26.5 ± 0.2 |
| 10 | 178 ± 25 | 460–581 | 766 ± 46 | 581–1101 | 25.5 ± 0.3 |
| 15 | 166 ± 63 | 438–573 | 720 ± 50 | 573–1046 | 26.4 ± 0.5 |
| 20 | 214 ± 18 | 434–571 | 700 ± 21 | 571–1096 | 25.8 ± 1.0 |

**Table 1.**
*Parameters from the results of the microwave-assisted dilatometry in high-frequency (30 GHz) curves as a function of heating rates for the nanometric hematite.*

According to the values in **Table 1**, there is a decrease in the initial and intermediate stage ranges can be observed, as well as a decrease in the maximum linear shrinkage temperature with the increase in the heating rate. Higher temperatures of onset of shrinkage can be observed with the increase in the heating rate. However, by the errors associated, no significant tendency can be stated. Except for the lowest heating rate (5°C/min) for maximum linear shrinkage rate, that is possible to state a higher temperature compared to the other heating rates.

These values were compared with the values reported by Togashi et al. [7], who used the same hematite nanopowder of this work and sintered it by conventional dilatometry. It is possible to observe an accentuated decrease in the onset of shrinkage temperature, using microwave energy at 30 GHz, and it varied approximately at 200° C, and with conventional heating approximately 700°C, indicating that the shrinkage process with the use of microwave at 30 GHz heating starts at lower temperatures. As it is highlighted in **Figure 3(b)**, this fact can be also attributed to the trials when realized by microwave dilatometry to produce a bigger temperature gradient between the trial material and the measurement system material, once the temperature distribution is highly dependent on dielectric properties of the analyzed material and on the materials surrounding it, as demonstrated by Link et al. [23]. Thus, for systems with relative high dielectric losses, such as the hematite, the interaction and electromagnetic field absorption is higher in the sample than in the alumina of the measurement system.

**Figure 4(a)** shows the graphical differences in the behavior of the linear shrinkage and the first derivative as a function of temperature in the microwave assisted in high-frequency (30 GHz) and conventional dilatometer measurements. **Figure 4(b)** shows the comparison between the microwave in high-frequency (30 GHz), microwave-assisted at 2.45 GHz (hybrid heating with the aid of a susceptor), and conventional dilatometer results, where the values at 2.45 GHz and conventional are from the literature and reported by Togashi et al. [7].

According to the literature [7], the final relative density of the hematite, of the same origin, after conventional dilatometry was a little lower as compared to hematite sintered at 30 GHz, with a value of approximately 87%, while the samples sintered at 2.45 GHz had a relative density value of 91%. However, as can be observed in **Figure 4**, the samples shrank more with the microwave assisted at 30 GHz heating,

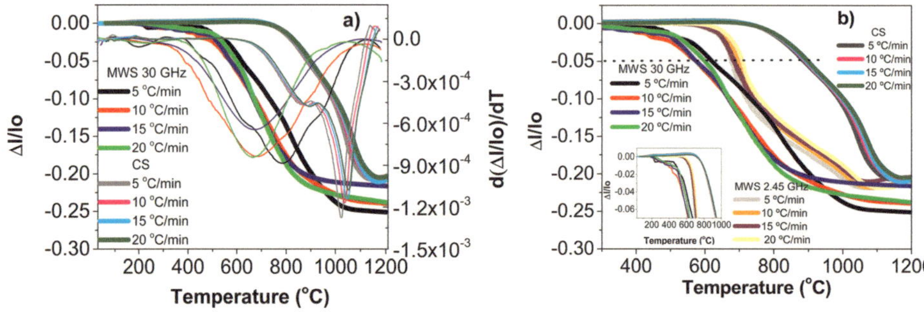


**Figure 4.**
*Dilatometric curves ($\Delta l/l_0$) and $d(\Delta l/l_0)/dT$ assisted by microwave in high frequency (30 GHz) and in conventional furnace from hematite nanopowder (a) and dilatometric curves ($\Delta l/l_0$) as a function of temperature microwave assisted at 30 GHz, in comparison with microwave at 2.45 GHz and in conventional furnace with the same hematite nanopowders [7] (b).*

showing a maximum linear shrinkage of 25.2%, while, according to the literature, in the conventional heating, this value was 20.8 and 21.8% at 2.45 GHz. The shrinkage is nonisotropic and a correction was made in the nonisothermal models.

Besides, in **Figure 4**, an accentuaded decrease in the envolved parameters is observed, such as: temperature of onset of shrinkage necessary for the microwave in high-frequency sintering, which is in the range of 200°C (*inset* **Figure 4(b)**), that is lower than with convetional heating, which was in the range of 673–676°C [7], to average reduction of approximately 500°C. In microwave at 2.45 GHz dilatometry this range was between 551 and 590°C, as reported in Ref. [7]; thus, there is an average reduction of 360°C in the temperature of onset of shrinkage with microwave sintering. The microwave at 2.45 GHz dilatometry has intermediates values, which are expected, since the susceptor used the sintering in this frequency was hybrid. Furthermore, theoretically in higher frequency (30 GHz) the absorbed power of the electromagnetic radiation by the material increases, increasing the contribution of the nonthermal effects during sintering. Link et al. [8] reported a reduction of 150°C in the sintering temperature, as well for the nanometric zirconia stabilized with yttria (36–37 nm) as for the PZT. The authors also reported a reduction in the dwell time from 60 to 10 min, when compared with the conventional sintering in order to obtain density values relatively similar. In this research, the reduction in the dwell time from 30 to 1 min was evaluated, and the results of densities were similar and of approximately 90% by microwave at 30 GHz heating.

There is a decrease in the beginning of the initial stage sintering temperature with the use of microwave in high-frequency heating compared with the literature data [7], the decrease in the microwave at 2.45 GHz heating is in the range of 100–185°C, while the decrease associated with the intermediate stage is in the range of 45–100°C. Compared with the conventional heating, this decrease is in the range of 270–365°C for the initial stage and 240–320°C for the intermediate stage of sintering. Once more, it was possible to observe lower temperatures necessary for the high-frequency (30 GHz) microwave sintering of the hematite. This is in accordance with what was published by Rybakov et al. [10], who reported a more drastic difference in the initial stage of sintering temperature, when compared with the intermediate and final stages of sintering of some ceramic materials, once the contribuition of the ponderomotive effect was higher [14].

An important difference to be highlighted in the comparison of the presented curves in **Figure 4(b)** is the intermediate behavior of the sintering kinetics at 2.45 GHz compared with 30 GHz and conventional sintering. In the temperature range of 740 and 800°C, close to the maximum shrinkage temperature, it is possible to observe a change in the slope of the linear shrinkage curve. At lower temperatures, the microwave assisted at 2.45 GHz dilatometry behaves similar to the conventional dilatometry. After the mentioned temperature range, the slope at 2.45 GHz is similar to the 30 GHz results. It is known that with hybrid heating when a susceptor material was used, the hematite behaves as a transparent material up to the range of 740–800°C and thereafter better absorbs the 2.45 GHz radiation, and temperature gradient between the hematite samples and the low loss alumina rod may occur. Thus, this temperature range corresponds to the so-called critical temperature, where predominant heating of the susceptor changes the predominant heating of the hematite sample [26]. The use of higher frequencies can eliminate the need of susceptor materials, since according to Eq. (1) the increase in frequency results in an increase in the absorbed power density as well as in the decrease in penetration depth (Eq. (2)) of the electromagnetic radiation.

In case of 30 GHz microwave sintering, the linear shrinkage curves slightly shift to lower temperature values with increasing heating rates. This can be explained by the volumetric heating effect going along with an inverse temperature profile and increasing temperature gradients with increasing heating rates [23]. However, this effect strongly depends on the dielectric losses of the material, since with higher dielectric losses and higher frequencies, and therefore lower penetration depth of the electromagnetic radiation, heating can be more superficial as well [6, 13], so that this trend in temperature shift can change. This is shown in **Figure 2(b)** for the hematite of this study, once there is low penetration depth and higher absorbed power.

In conventional heating where the shrinkage curves shift to higher temperatures, the opposite behavior can be seen. This is due to the delay in heat transfer because of low thermal conductivity of the powder compact, as also verified by Mazaheri et al. [27] for the conventional heating of zirconia nanopowder. Thus, the inversion in the tendency observed for the microwave in high frequency is possible to be associated with the increase in the diffusion mechanisms rate or a contribution of the inversion of temperature profiles. This contributes both to densification as grain growth for the ceramic materials. In case of the 2.45 GHz hybrid heating results, the trend with increasing heating rates is more comparable to conventional heating, due to the use of susceptor.

### 3.2 Densification kinetics

Obviously, only the decrease in the analyzed temperature parameters in the dilatometric trials, in an isolated form, cannot be the only resource in order to analyze the microwave-assisted sintering. Thus, for a better understanding of the densification process of a ceramic material during sintering, kinetics studies were performed, with approximate classic models, which are applied in submicron ceramic systems and conventional heating, and they were adapted to nanometric systems and microwave-assisted heating. All the trials were made in triplicate. And the temperature measurement was performed with a thermocouple, carefully placed in the sample. Once wrong temperature measurement could lead to wrong interpretation of the models.

The initial stage of the microwave-assisted sintering at 30 GHz was analyzed through the application of non-isothermal methods. Eq. (4) from Woolfrey-Bannister [24] shows the relationship between the product of temperature T and time derivative of the linear shrinkage *dY/dt*, and the linear shrinkage, $Y = \Delta l/lo$ with activation energy *Q* for densification in the initial stage of sintering

$$T^2 \frac{dY}{dt} = \frac{Q}{(n+1)R} Y \tag{4}$$

where *n* is a characteristic factor for dominant sintering mechanism and $R = 8.31$ J/mol.K is gas constant.

In order to use Eq. (4), Eq. (5) from the Dorn model is needed, where the activation energy is calculated from the instantaneous shrinkage derivative as a function of temperature *V1* and *V2*, where V = dY/dT, referent to the temperatures *T*1 and *T*2, which delimitate the start and end of the initial stage of sintering.

$$Q \approx \frac{RT1T2}{T1 - T2} ln \frac{V1}{V2} \tag{5}$$

The initial stage of sintering is determined as the linear region of the graphic from Eq. (4), and they were obtained by the curve in **Figure 5**, as schematized below. Up to 1% there is a pre-initial stage of sintering where there is only the approach of the particles, but no significant shrinkage. **Figure 5** highlights the region of the initial stage, and in detail, it is possible to see the entire graphic. The activation energy is calculated by Eq. (5), and in this case, with the slope of the curve in **Figure 5**, it is possible to calculate n, the dominant sintering mechanism. However, the value near to 0 indicates the nanoparticle rearrangement, and not the mechanism itself. In order to corroborate the range of initial stage sintering, here estimated between 1 and 4%, the graphic of LndY/dt vs. Y from Johnson model is presented. The linear region in the same range of **Figures 5** and **6** only corroborates the estimation of the initial stage of sintering.

The Johnson model considers the two spheres model, and the initial stage is analyzed by constant heating rate. Eq. (6) correlates the linear shrinkage rate (dY/dt) with particle size (g) and temperature (T) [28].

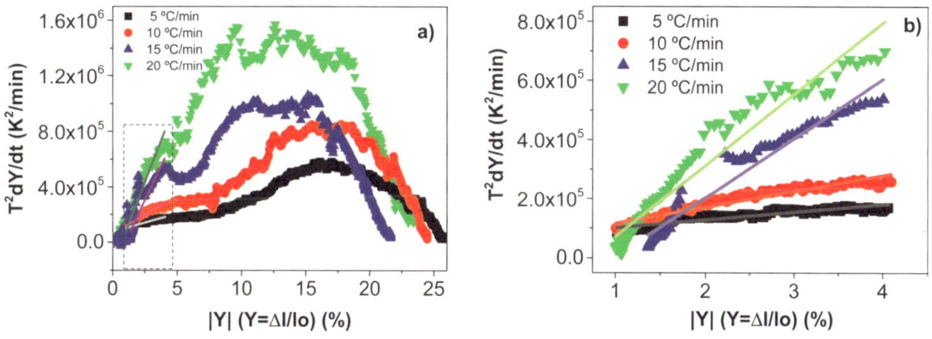


**Figure 5.**
*Graphics from Woolfrey and Bannister model for the determination of the initial stage having* $T^2dY/dt$ *as a function of |Y| for different heating rates (a) and zoom linear fitting of the initial stage of sintering estimated between 1 and 4% of linear shrinkage (b).*

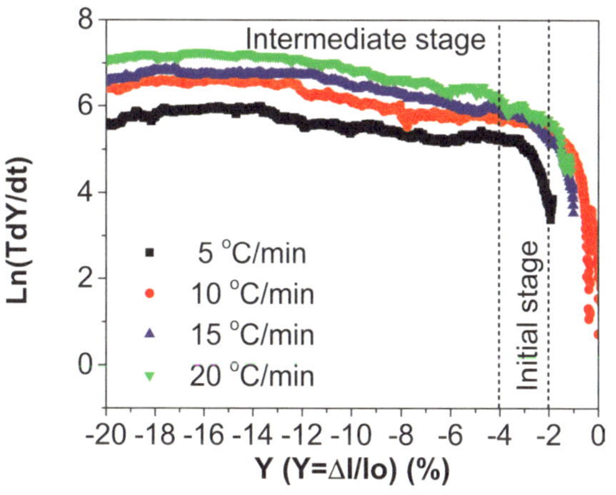


**Figure 6.**
*Graphics from the Johnson model* Ln(TdY/dt) *as a function of* Y *with different heating rates.*

$$\frac{dY}{dt} = A_0 \frac{1}{g^{m_1} Y^{m_2}} \frac{e^{\left(-\frac{Q}{RT}\right)}}{T} \tag{6}$$

$A_0$ is a constant depending on the material and the sintering mechanism, $m_1$ and $m_2$ are exponents which have values of $m_1 = 1$ and $m_2 = 0$ for viscous flow, $m_1 = 3$ and $m_2 = 1$ for bulk diffusion, and $m_1 = 4$ and $m_2 = 2$ for grain boundary diffusion.

$Y = \Delta l/lo$ is the linear shrinkage in percentage, $Q$ is the activation energy, and R is the gas constant. Adjusting and applying logarithm in both sides, Eq. (7) is obtained.

$$\ln\left(T\frac{dY}{dt}\right) = -\frac{Q}{R}\frac{1}{T} + \ln A_0 - m_1 \ln g - m_2 \ln Y \tag{7}$$

**Figure 6** shows the graphic from Johnson model, corroborating a linear region of the graphic in the range of approximately 1 to 4%, corresponding to the initial stage of sintering.

According to **Figures 5** and **6**, the linear region corresponds to the initial stage, and it can be estimated, in this case, occurring from 1 to 4% of the linear shrinkage. In the literature, the initial stage is predicted when the densification reaches a maximum of 60–65%, depending on the studied material and green density [29]. Woolfrey [30] studied $UO_2$ and estimated the initial stage between 600 and 1200°C in $H_2$ atmosphere. The author concluded that with the decrease in green density, below the so-called critical range, there is a progressive decrease in the sintering rate. However, the green density does not affect the apparent activation energy values in the initial stage, despite it has an important effect in the intermediate stage.

In this study with hematite, starting with a nanometric powder, the analysis of the curves allowed the estimation of the initial stage up to 45% of the instantaneous density, corresponding to approximately 4% of linear shrinkage. This value was assumed, in view of the theory and together with the insightful analyses of **Figure 5**, knowing that the initial stage corresponds to the straight line of the graphic $T^2dY/dt$ vs. $Y$. Thus, in this case, the initial stage for the studied material ends at 45% of density, where begins immediately the intermediate stage.

The densification curve is calculated by Eq. (8):

$$\rho = \frac{\rho_0}{\left(1 + \frac{\Delta l}{l_0}\right)^3} \tag{8}$$

The instantaneous density, ρ, of the samples in function of the shrinkage is given by Eq. (8), where $\rho_0$ is the green density, $l_0$ is the initial length of the sample and $\Delta l$ is the linear shrinkage, where $\Delta l = l - l_0$ [31].

The instantaneous density values calculated by Eq. (8) are expected to be slightly smaller than the apparent density values calculated by Archimedes' principle, as shown in Table IV later, since instantaneous density is an approximate statistical calculation, which considers only the linear shrinkage of the sample. While in the principle of Archimedes, the dry, humid, and immersed masses are estimated, considering the porosity of the sample. **Figure 7** presents the densification graphic with the increase in temperature, depending on heating rate. In this case, it used the relative density, knowing that the theoretical density for hematite is 5.24 g/cm$^3$.

**Table 2** shows the obtained values of the activation energy for the densification in the microwave at 30 GHz initial stage of sintering in comparison to corresponding

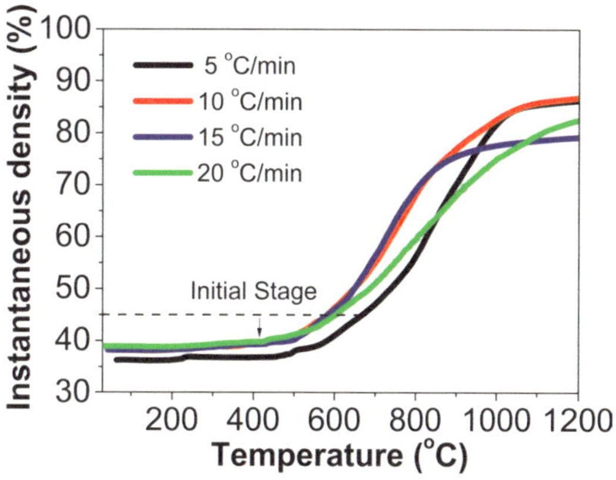


**Figure 7.**
*Graphic of instantaneous density vs. temperature with heating rates of 5 to 20°C/min.*

results obtained with 2.45 GHz [7] and conventional heating. These values were calculated by Eq. (5) after the estimation of the temperatures and shrinkages related to the begin and end of initial stage of sintering, as presented in the third row of **Table 1**.

As can be seen in **Table 2** the activation energy obtained for initial stage sintering at 30 GHz decreases with increasing in heating rates. Furthermore, the values are significantly lower obtained for conventional and 2.45 GHz microwave sintering [7]. For conventional heating, the calculated range of activation energy for densification in the initial stage of sintering was between 179 and 168 kJ/mol, following the same tendency of decrease in energy values with the increase in heating rate. The microwave-assisted sintering at 2.45 GHz shows intermediate values in the range of 161–151 kJ/mol. Thus, a reduction of approximately 74 to 60%, compared with 2.45 GHz and 77 to 63%, with conventional, in the needed energy is observed for the densification in the initial stage.

The driving force for the sintering process is the decrease in free energy, by the reduction of energy associated with surfaces, and the process occurs by diffusion mechanisms and mass transport. It is reasonable to assume that these mechanisms are accelerated when using microwave radiation as an additional driving force, which would reduce the associated temperatures, as well as the energy for the initial stage of densification.

| Heating rate (°C/min) | Q (Initial stage) (kJ/mol) | | |
|---|---|---|---|
| | Microwave at 30 GHz | Microwave at 2.45 GHz [7] | Conventional [7] |
| 5 | 66 ± 3 | 161 ± 4 | 179 ± 6 |
| 10 | 48 ± 9 | 160 ± 3 | 175 ± 6 |
| 15 | 47 ± 2 | 156 ± 4 | 174 ± 9 |
| 20 | 39 ± 5 | 151 ± 3 | 168 ± 4 |

**Table 2.**
*Activation energy for the initial stage of sintering by microwave at 30 GHz of nanometric hematite compared with the results of the same material in Ref. [7].*

On the other hand, for the analysis of the intermediate stage of sintering the heating rate model Wang and Raj [25] was used. The authors estimated the intermediate stage range between 65 and 85% for a green density equal to 55%. Zhu et al. [2] estimated the intermediate stage between 70 and 90%, with an green density of approximately 48%. For this work, the intermediate stage range was identified between 45 and 85% of relative density. This estimation was made observing the end of the initial stage 4% of linear shrinkage and, in this case, corresponds to approximatedely 45% of relative density, as well as the observation of the curves $\ln\left(\frac{Td\rho}{dT}\right) \times \frac{10^{-4}}{T}$ in **Figure 8**. The intermediate stage corresponds to the straight lines of these graphics, which has inclinations with near values. Thus, by the cited method, the densification rate *dρ/dT* given by Eq. (9) depends on factors such as dimensionless constant *A*, the activation energy for densification in the intermediate stage *Ea*, a function *f(p)* which depends only on the density, the grain size G, and the dominant mechanism of sintering *m* and the heating rate *dT/dt*.

$$\frac{d\rho}{dT} = A\frac{e^{\frac{-Ea}{RT}}}{T}\frac{f(p)}{g^m}\left(\frac{dT}{dt}\right)^{-1} \tag{9}$$

Eq. (8) can be written as Eq. (10):

$$ln\left(T\frac{d\rho}{dT}\right) = -\frac{Ea}{RT} + ln\left(f(p)\right) + ln\ A - m\ ln\ g - ln\left(\frac{dT}{dt}\right) \tag{10}$$

It is assumed that after sintering, the grain size *g* depends solely on the density, and thus, through an Arrhenius plot of the densification as a function of inverse temperature, at the different heating rates, the activation energy for densification *Ea* for intermediate stage of sintering can be estimated. The mentioned Arrhenius plot is presented in **Figure 8** for 30 GHz sintering and for relative densities in the range from 45 to 85%, which corresponds to intermediate stage.

And by Eq. (10), the activation energy for the intermediate stage of microwave-assisted sintering at 30 GHz of the nanometric (30 nm) hematite was calculated to

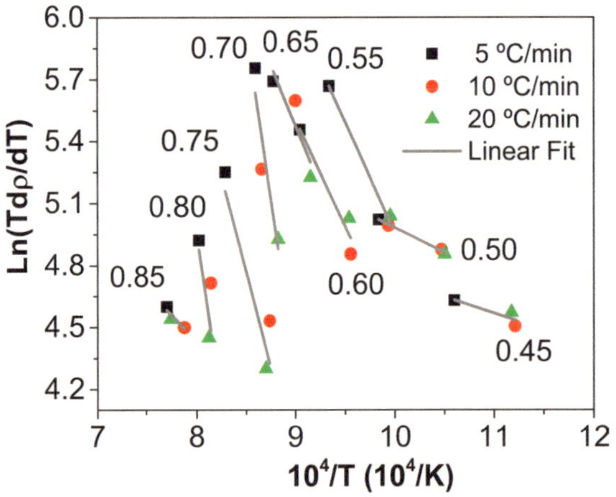


**Figure 8.**
*Arrhenius plot of Ln (Tdρ/dT) as a function of the inverse of temperature in the range of 45 to 85% of relative density, corresponding to the intermediate stage of microwave at 30 GHz sintering of hematite.*

68 ± 15 kJ/mol. According to the literature [7], the activation energy for the intermediate stage using the microwave-assisted heating at 2.45 GHz dilatometric data was 273.1 ± 28.0 kJ/mol and for the conventional heating was 485.0 ± 34.0 kJ/mol for the same material. It is possible to observe a reduction of approximately 70 and 83% in the necessary energy for densification, when compared with microwave-assisted at 2.45 GHz and conventional heating, respectively, similar to the results obtained for the initial stage of microwave at 30 GHz sintering.

The microwave heating at 30 GHz of frequency, which has the electromagnetic wave in the millimetric range, can raise the nonthermal effect, reducing the temperatures compared with the conventional heating, besides the reduction observed in the densification kinetics. This nonthermal effect possibly generates the increase in the mass transport and diffusion, of thermally activated processes, such as sintering, which can be explained by the action of the ponderomotive force [4, 12]. Considering that these forces depend on mobile vacancies in ionic crystalline solids, as ceramic materials, they carry effective electrical charges which move and oscillate in the applied frequency; in this way, in higher frequencies, there is the increase in the processes of mass transport in ionic crystalline solids and the presence of grain boundaries near the pores act as vacancies sources, increasing the diffusivity in this region, and the nanometric materials potentially increase also the diffusivity compared to submicrometric materials [12, 32]. And this increase of mass transport, which results from the ponderomotive effect, is caused by an additional driving force that influences the reaction kinetics of sintering. However, the influence of the microwave radiation action decreases with the time of process, as a consequence of the changes in the microstructure of the sample itself, such as the neck formation and open porosity closure [14, 15].

### 3.3 *In situ* electrical resistivity

Parallel to dilatometric measurements, *in situ* electrical resistivity measurements were performed during 30 GHz sintering for nanometric (30 nm) hematite samples as a function of temperature, and with a heating rate of 20°C/min. This allows to evaluate the effect of densification on the electrical resistivity of the material. Values of $\Delta l/lo$, log ρ and relative density (according to Eq. (8)) as a function of temperature are shown in **Figure** 7. With the results of the first analysis of the heating cycle in the green sample (first cycle) in **Figure 9(a)**, the second analysis referring to the sintered sample (second cycle) in **Figure 9(b)**.

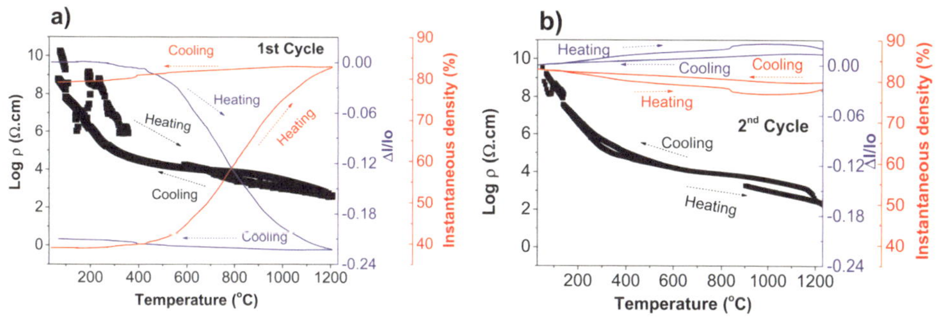


**Figure 9.**
In situ *electrical resistivity of nanometric hematite concurrent to microwave at 30 GHz dilatometry at 20°C/min (a) first cycle and (b) second cycle of heating.*

| | First cycle | Second cycle |
|---|---|---|
| $P_{max}$ [Ω.cm] | $1.7\times10^{10}$ | $6.0\times10^{8}$ |
| $P_{min}$ [Ω.cm] | $3.8\times10^{2}$ | $4.3\times10^{2}$ |

**Table 3.**
*Maximum and minimum electrical resistivity measured during trial* in situ *of hematite in microwave at 30 GHz.*

**Table 3** presents the maximum and minimum electrical resistivity measured in these trials, according to Eq. (3).

The *in situ* electrical resistivity values measured were in the range of $3.8\times10^2$ to $1.7\times10^{10}$ Ω.cm in the first cycle of heating and between $4.3\times10^2$ and $6.0\times10^8$ Ω.cm in the second cycle of heating. It is possible to observe in the first cycle of heating, which refers to the densification cycle of the sample, the decrease in electrical resistivity values was in function of the increase in the relative density (porosity reduction). In the second cycle of heating (annealing), the sintered sample has a density around 85% and lower values of electrical resistivity in temperatures are observed below 200°C, as expected. Thus, the direct relationship between the material electrical behavior and density of the sample was observed.

It is known that the electrical conductivity is an intrinsic property of the material related to the facility of electrons transport; thus, the resistivity indicates the imposed resistance to this electronic movement [33]. In this sense, Kultayeva et al. [34] studied SiC containing alumina as a sintering aid and concluded that with the decrease in porosity, the electrical conductivity of the SiC was improved. This property depends strongly on microstructure and porosity of the material.

The results obtained and showed in the second cycle of heating from **Figure 9(b)** demonstrate the semiconductor behavior of hematite, by the reduction in the values of electrical resistivity as a function of the increase in temperature, by the greater diffusion energy from the charge carries with the increase in temperature.

Others works, such as Link et al. [16], studied the metalic compacts sintering (pure iron and with graphite) using the same measurement system of *in situ* electrical resistivity as described in this work. The authors observed lower values of resistivity at 300°C, corresponding to the interparticle contact, as in the range of 600 to 700°C, due to the metalic powders starting to sinter. Tian-Ming et al. [35] studied the effect of porosity on electrical resistivity of carbon-based materials for electrode applications, and the authors demonstrated that both open and closed porosities increase the electrical resistivity values. However, the open porosity is the one that has the greater influence in the values of electrical properties of the material. **Figure 10** presents the Arrhenius plot, considering a graphic of Log Tσ vs. $10^4$/T.

The dependence of the electrical conductivity (σ) with the temperature (*T*) is expressed by Eq. (11).

$$\sigma = \frac{\sigma_0}{T} e^{-\frac{E_{at}}{kT}} \tag{11}$$

$\sigma_0$ is a pre-exponential factor related with conductivity, $E_{at}$, the activation energy of the conduction process and k the Boltzmann constant, and its value is $13{,}806\times10^{-22}$ J/K or $86{,}177\times10^{-4}$ eV/K.

Thus, applying logarithm for both sides, Eq. (12) is obtained.

$$\log T\sigma = \log \sigma_0 - \frac{E_{at}}{kT} \tag{12}$$

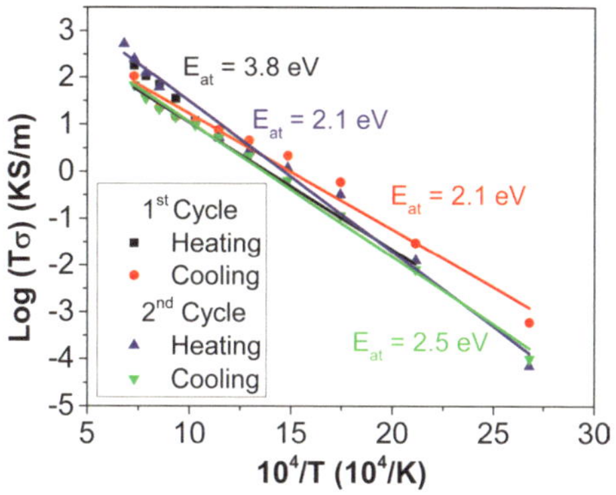


**Figure 10.**
*Arrhenius plot of log $T\sigma$* vs. *$10^4/T$ during heating and cooling in the first and second cycles.*

This activation energy, differently of the activation energies previously presented, represents the needed energy in order to the electrons move inside the structure of the material [36]. The activation energy, in this case, decreases in the first cycle from heating to cooling. Going from 3.8 to 2.1 eV. The energy increases in the second cycle in the cooling process to 2.5 eV. Normally, lower values of activation energy are related to higher electrical conductivity values, once the energy needed for electron movement would be lower, generating thus, better mobility [36]. The decrease of the value of activation energy in the first cycle is coherent with the fact that densification is happening during sintering, and the conductivity increases during heating.

### 3.4 Final microstructure

For the microstructural evaluation, the samples sintered at 30 GHz with the heating rates of 5 and 20°C/min were sectioned in half of its length, as described in the scheme above **Figure 11**, to evaluate the heating homogeneity. Each section was carefully grinded and polished and analyzed using optical microcopy. The results are presented in **Figure 11**.

The optical micrographs in **Figure 10** show a slight change in open porosity distribution. According to the evaluation with ImageJ program, in the threshold analysis, the percentage of open porosity is approximately 36–42% for (b-d) image and 30–28% for (a-c) images, respectively. So, it is possible to conclude that on the surface (b-d), the porosity is higher than in the center (a-c). This may indicate a thermal gradient between the surface and the center of the sample, which is expected in microwave sintering. The presence of cracks or flaws through the sample volume was not observed, indicating that there is no differential shrinkage.

The final microstructure analyzed by SEM images of the samples submitted to microwave assisted dilatometry at 30 GHz in the different heating rates is presented in **Figure 12**.

**Figure 12** analysis shows the SEM micrographs and the final grain size distribution as a function of heating rate from 5 to 20°C/min. The reduction of the average grain sizes at higher heating rates can be observed. The final grain size distribution tends to be narrower at higher heating rates, as presented in **Figure 13**. The grain size distribution histograms, as represented in **Figure 13**, were fitted into a LogNormal curve distribution, according to Eq. (13).

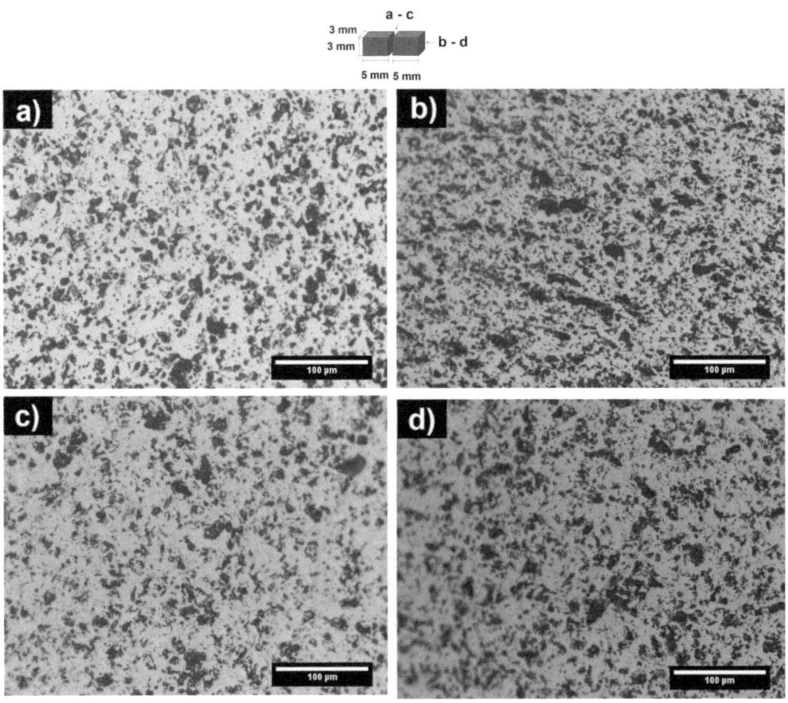


**Figure 11.**
*Optical micrographs of hematite after sintering at 30 GHz with a final temperature of 1200°C, dwell time of 1 minute, and heating rate of 5°C/min (a-b) and 20°C/min (c-d).*

$$y = y_0 + \frac{A}{\omega'\sqrt{2\pi}} e^{\left(-\frac{\left(\ln\frac{x}{x_c}\right)^2}{2\omega^2}\right)} \tag{13}$$

$y$ and $y_0$ are free parameters, and $y_0$ tends to zero, $A$ is an amplitude factor, $\omega'$ the width of the peak associated to the dispersion, and $x_c$ the central point of the peak, corresponding to the average grain diameters. The values of $\omega'$ in the different heating rates were: 0.097, 0.085, 0.084, and 0.047 for 5, 10, 15, and 20°C/min, which indicates with the narrower grain size distribution with higher heating rates.

Furthermore, the values of the average grain size are presented in **Table 4** with the final relative density values, calculated by the principle of Archimedes.

This tendency of lower grain sizes and lower densities with the increase in heating rate was expected, as reported in the literature by Chu et al. [37] who evaluated this kind of behavior of ZnO under conventional sintering, where lower grain sizes were observed at higher heating rates. Bykov et al. [38] observed the sintering kinetics by the isothermal method and obtained lower grain sizes using millimeter wave sintering compared to the conventional sintering.

Lange [39] studied the conventional sintering of alumina and postulated that the maximum linear shrinkage rate occurred with a relative density of 77%. The author observed the alteration in sintering kinetics. Until the temperature associated with this density, the densification processes dominate, and after that, the grain growth happens. For the hematite analyzed in this work, this maximum shrinkage rate temperature occurs in the range of densities of 65%. Thus, from this densification higher grain growth is expected, even in case of high-frequency microwave sintering.

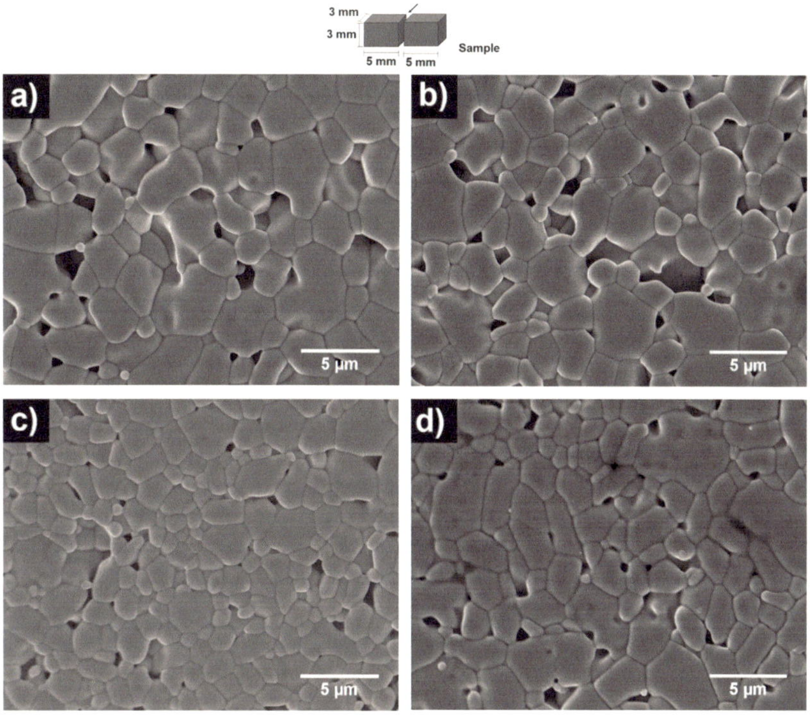


**Figure 12.**
*SEM images of hematite after 30 GHz microwave sintering at 1200°C, dwell time of 1 minute, and heating rate of (a) 5°C/min, (b) 10°C/min, (c) 15°C/min, and (d) 20°C/min.*

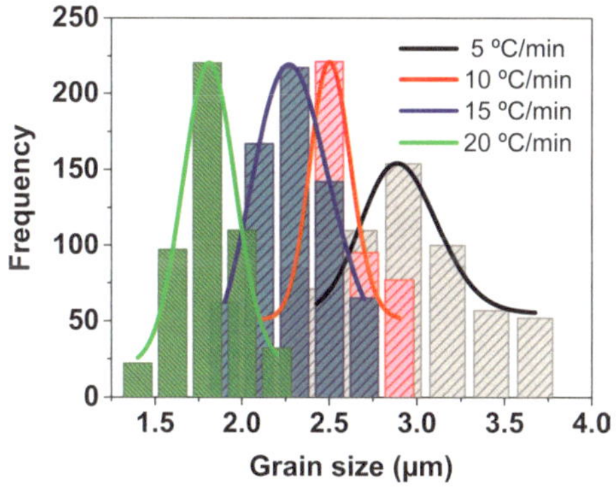


**Figure 13.**
*Comparative histograms from the average grain sizes with different heating rates.*

| Heating rate (°C/min) | Average grain size (µm) | Relative density (%) |
|---|---|---|
| 5 | 2.8 ± 0.4 | 88 ± 2 |
| 10 | 2.5 ± 0.3 | 87 ± 1 |
| 15 | 2.3 ± 0.2 | 86 ± 3 |
| 20 | 1.9 ± 0.2 | 84 ± 3 |

**Table 4.**
*Average grain size and final relative density of hematite after at 30 GHz microwave sintering.*

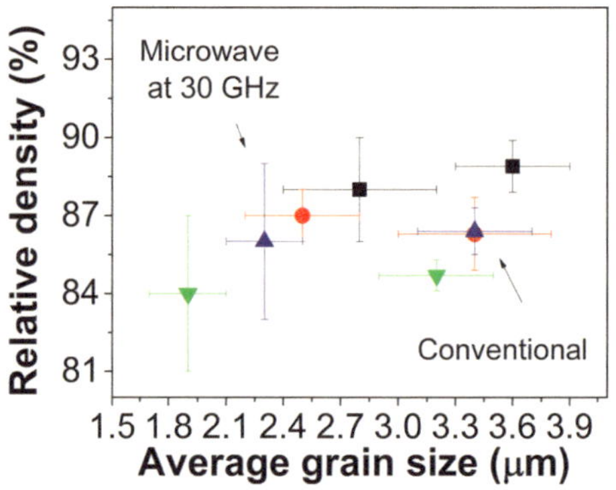


**Figure 14.**
*Relative density as a function of the average grain sizes of hematite sintered by microwave at 30 GHz and conventional [7] at 1200°C in the different heating rates.*

The average grain sizes in the investigated range of heating rates were in the range from 1.9 to 2.8 μm for 30 GHz microwave sintering. In case of conventional sintering of identical hematite powder, the range was 3.2 to 3.6 μm [7]. **Figure 14** shows the comparison between these values in function of apparent relative density. There is a clear effect of heating rate on the final grain size and relative final density for both microwave heating at 30 GHz and conventional heating [7].

Higher density values were observed at lower heating rates resulting in larger grain sizes after sintering. This observation is consistent with the fact that during sintering, there is a strong competition between densification and grain growth, due to the fact that both processes have the same driving force, which is the decrease in the free energy of the system, and specifically in this case, the decrease in the free energy associated with surfaces [40].

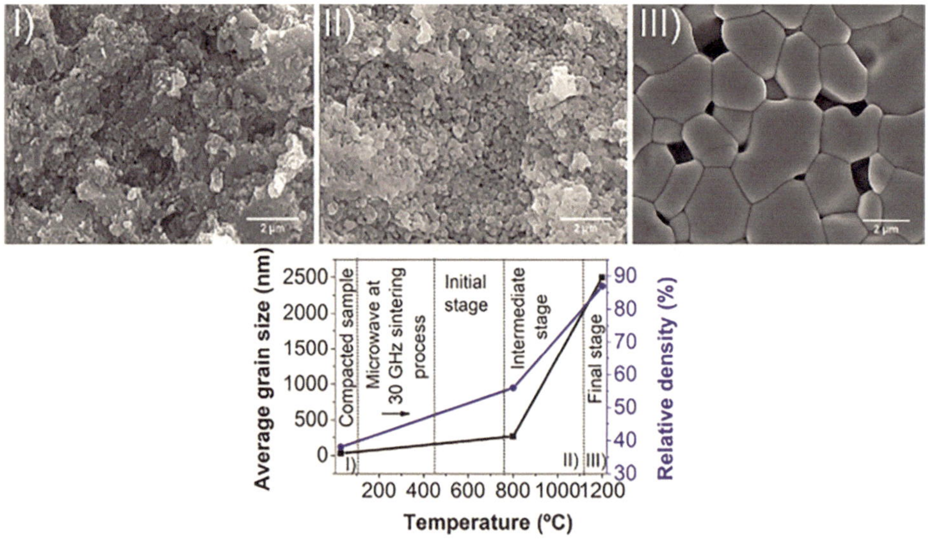


**Figure 15.**
*Evolution of the grain size, density, and microstructure of the compacted hematite (30 nm) during microwave sintering at 30 GHz and heating rate of 10°C/min.*

The evolution in grain size, density and microstructure, starting from the nanometric hematite compact, until reaching the beginning of the intermediate stage, near to the temperature of 800°C and up to the final stage obtained at 1200°C is presented in **Figure 15**. The micrographs refer to microwave sintering at 30 GHz performed with a heating rate of 10°C/min.

From the microstructural evolution presented in **Figure 15**, it can be observed that the intermediate stage of sintering is responsible for pore rounding, neck formation that show that the beginning of the diffusion mechanisms favors the grain growth, as well as the densification during microwave-assisted sintering at high frequency (30 GHz). The average grain size with a heating cycle up to 800°C, where the intermediate sintering stage started, was 270 ± 30 nm with a grain growth of approximately 90% in relation to the initial size of the nanopowders of 30 nm. The final bulk density in this condition was approximately 56% measured by the immersion method. The heating rate was 10°C/min. At the end of the process, the obtained grain size was 2.5 ± 0.3 μm at a relative density of about 87%

## 4. Conclusions

The sintering kinetics models from nonisothermal methods, such as the classics by Woolfrey and Bannister and by Dorn for the initial stage of sintering, and by Wang and Raj, for the intermediate stage, have been applied to *in situ* dilatometer measurement results during 30 GHz microwave sintering of hematite nanopowders with heating rates varying from 5 to 20°C. Densification to relative densities close to 90% was achieved, regardless of the heating rate. The results demonstrated the effectiveness of using these models.

The results showed that with all sintering conditions, the onset of shrinkage as well as the ranges of initial and intermediate stages of sintering was observed at lower temperatures when compared to 2.45 GHz microwave sintering as well as conventional sintering of the same material. Based on these models, the activation energies for the initial stage and for the intermediate stage of sintering have been estimated to be 39–66 kJ/mol and 68 kJ/mol, respectively. Compared to 2.45 GHz and conventional sintering, those activation energies were significantly lower, which might be an indication of nonthermal effects due to additional driving forces such as the ponderomotive force.

Less grain growth was observed compared to conventional heating. For microwave sintering at 30 GHz grain sizes between 1.9 and 2.8 μm were obtained. It was also observed that with increasing heating rates the average grain size was smaller and has narrower grain size distribution. Moreover, in all cases, the microstructures were homogeneous, showing controlled and uniform heating.

From the *in situ* electrical resistance measurements, the decrease in resistivity as a function of the increase in the density of the sample could be observed, as well as a semiconductor behavior of the hematite.

## Acknowledgements

The authors gratefully acknowledge the financial support of the Brazilian research funding agencies: FAPESP (Process no. 2017/13769-1), CAPES (Process 88882.332729/2019-01 and 88887.370181/2019-00), CNPq (Process 305129/2018-0 and 165313/2017-0). This study was financed in part by the Coordenação de Aperfeiçoamento de Pessoal de Nível Superior – Brasil (CAPES) – finance code 001.

## Author details

Marina Magro Togashi[1], Claudia P. Fernandez Perdomo[1*], Guido Link[2], Jhon Jelonnek[2] and Ruth Herta G.A. Kiminami[1]

1 Graduate Program in Materials Science and Engineering, Materials Science Department, Federal University of São Carlos, São Carlos, Brazil

2 Karlsruhe Institute of Technology (KIT), Institute for Pulsed Power and Microwave Technology (IHM), Germany

*Address all correspondence to: claudia.perdomo@ufscar.br

## References

[1] Agrawal D. Microwave sintering of ceramics, composites and metal powders. In: Fang Z, editor. Sinter. Adv. Mater. Fundam. Process. Duxford, Kidlington, United Kingdom and Cambridge, United States of America: Woodhead Publishing; 2010. pp. 222-248

[2] Zhu J, Ouyang C, Xiao S, Gao Y. Microwave sintering versus conventional sintering of NiCuZn ferrites. Part I: Densification evolution. Journal of Magnetism and Magnetic Materials. 2016;**407**:308-313

[3] Zuo F, Badev A, Saunier S, Goeuriot D, Heuguet R, Marinel S. Microwave versus conventional sintering: Estimate of the apparent activation energy for densification of α-alumina and zinc oxide. Journal of the European Ceramic Society. 2014;**34**: 3013-3110

[4] Zuo F, Saunier S, Marinel S, Chanin-Lambert P, Peillon N, Goeuriot D. Investigation of the mechanism(s) controlling microwave sintering of α-alumina: Influence of the powder parameters on the grain growth, thermodynamics and densification kinetics. Journal of the European Ceramic Society. 2015;**35**:959-970

[5] Brosnan KH, Messing GL, Agrawal DK. Microwave sintering of Alumina at 2.45 GHz. Journal of the American Ceramic Society. 2003;**86**: 1307-1312

[6] Sudiana IN, Mitsudo S, Inagaki S, Ngkoimani LO, Rianse U, Aripin H. Apparent activation energy in high frequency microwave sintering of alumina ceramic. Advanced Materials Research. 2015;**1123**:391-396

[7] Togashi MM, Perdomo CPF, Kiminami RHGA. Densification kinetics of nano-hematite using microwave assisted dilatometry. Ceramics International. 2020;**46**:28546-28560

[8] Link G, Rhee S, Feher L, Thumm M. Millimeter wave sintering of ceramics. In: Heinrich JG, Aldinger F, editors. Ceram. Mater. Components Engines. Germany: Wiley-VCH Verlag GmbH, Winheim; 2001. pp. 457-461

[9] Thumm M, Feher L, Link G. Micro- and millimeter-wave processing of advanced materials at Karlsruhe research center. In: Nov. Mater. Process. By Adv. Electromagn. Energy Sources, Proceedings of the International Symposium on Novel Materials. Processing by Advanced Electromagnetic Energy Sources March 19–22, 2004, Osaka, Japan. 2005. pp. 93-98

[10] Rybakov KI, Olevsky EA, Krikun EV. Microwave sintering: Fundamentals and Modeling. Journal of the American Ceramic Society. 2013;**96**: 1003-1020

[11] Paul F, Menesklou W, Link G, Zhou X, Haußelt J, Binder JR. Impact of microwave sintering on dielectric properties of screen printed $Ba_{0.6}Sr_{0.4}TiO_3$ thick films. Journal of the European Ceramic Society. 2014;**34**: 687-694

[12] Birnboim A, Gershon D, Calame J, Birman A, Carmel Y, Rodgers J, et al. Comparative study of microwave sintering of zinc oxide at 2.45, 30, and 83 GHz. Journal of the American Ceramic Society. 1998;**81**:1493, 1501

[13] Sudiana IN, Ito R, Inagaki S, Kuwayama K, Sako K, Mitsudo S. Densification of alumina ceramics

sintered by using submillimeter wave gyrotron. Journal of Infrared, Millimeter, and Terahertz Waves. 2013;**34**:627-638

[14] Rybakov KI, Olevsky EA, Semenov VE. The microwave ponderomotive effect on ceramic sintering. Scripta Materialia. 2012;**66**: 1049-1052

[15] Rybakov KI, Semenov VE, Link G, Thumm M. Preferred orientation of pores in ceramics under heating by a linearly polarized microwave field. Journal of Applied Physics. 2007; **101**(084915):1-5

[16] Link G, Mahmoud MM, Thumm M. Dilatometric study and in situ resistivity measurements during Millimeter wave sintering of metal powder compacts. Processing and properties of advanced ceramics and composites. IV Ceramic Transactions. 2012;**234**:145-149

[17] Pomar CD, Martinho H, Ferreira FF, Goia TS, Rodas ACD, Santos SF, et al. Synthesis of magnetic microtubes decorated with nanowires and cells. AIP Advances. 2018;**8**:045008

[18] Li J, Chu D. Energy band engineering of metal oxide for enhanced visible light absorption. In: Lin Z, Ye M, Wang M, editors. Multifunctional Photocatalytic Materials for Energy. Duxford, Kidlington, United Kingdom and Cambridge, United States of America: Woodhead Publishing; 2018. pp. 49-78

[19] Muhajir M, Puspitasari P, Razak JA. Synthesis and applications of hematite $\alpha$-$Fe_2O_3$: A review. Journal of Mechanical Science and Technology. 2019;**3**:51-58

[20] Ramya SIS, Mahadevan CK. Effect of calcination on the electrical properties and quantum Cofinement of $Fe_2O_3$ Nanoparitcles. International Journal of Research in Engenering and Technology. 2014;**3**:570-581

[21] Soldatov S, Umminger M, Heinzel A, Link G, Lepers B, Jelonnek J. Dielectric characterization of concrete at high temperatures. Cement and Concrete Composites. 2016;**73**:54-61

[22] Metaxas AC. Foundations of Electroheat, a Unified Approach. 1st ed. Chichester, United Kingdom: John Wiley & Sons; 1996

[23] Link G, Rhee S, Thumm M. Dilatometer measurements in a mm-wave oven. In: Willert-Porada M, editor. Advances in Microwave and Radio Frequency Processing. Berlin, Heidelberg, Germany: Springer; 2006. pp. 506-513

[24] Woolfrey JL, Bannister MJ. Nonisothermal techniques for studying initial-stage sintering. Journal of the American Ceramic Society. 1972;**55**: 390-394

[25] Wang J, Raj R. Estimate of the activation energies for boundary diffusion from rate-controlled sintering of pure alumina, and alumina doped with zirconia or Titania. Journal of the American Ceramic Society. 1990;**73**(5): 1172–1175

[26] Menezes RR, Souto PM, Kiminami RHGA. Microwave sintering of ceramic materials. In: Lakshmanan A, editor. Sinter. Ceram. - New Emerg. Tech. London, UK, London: Intech Open; 2012. pp. 3-25

[27] Mazaheri M, Simchi A, Dourandish M, Golestani-Fard F. Master sintering curves of a nanoscale 3Y-TZP powder compacts. Ceramics International. 2009;**35**:547-554

[28] Johnson DL. New method of obtaining volume, grain-boundary, and surface diffusion coefficients from sintering data. Journal of Applied Physics. 1969;**40**:192-200

[29] Barsoum MW. Fundamentals of Ceramics. 2nd ed. Boca Raton, United States of America: CRC Press; 2019

[30] Woolfrey JL. Effect of green density on the initial-stage sintering kinetics of $UO_2$. Journal of the American Ceramic Society. 1972;**55**:383-389

[31] Rahaman MN. Ceramic Processing and Sintering. 2nd ed. Florida: Taylor and Francis group; 2003

[32] Rybakov KI, Semenov VE. Mass transport in ionic crystals induced by the ponderomotive action of a high-frequency electric field. Physical Review B. 1995;**52**:3030-3033. DOI: 10.1103/physrevb.52.3030 accessed February 15, 2023

[33] Heaney MB. Electrical conductivity and resistivity. In: Webster JG, editor. Electrical Measurement, Signal Processing, and Displays. Boca Raton, United States of America: CRC Press; 2003

[34] Kultayeva S, Ha JH, Malik R, Kim YW, Kim KJ. Effects of porosity on electrical and thermal conductivities of porous SiC ceramics. Journal of the European Ceramic Society. 2020;**40**: 996-1004

[35] Tian-ming S, Li-min D, Chen W, Wen-Li GUO, Li W, Tong-xiang L. Effect of porosity on the electrical resistivity of carbon materials. New Carbon Materials. 2013;**28**:349-354

[36] Sharma M, Yashonath S. Correlation between conductivity or diffusivity and activation energy in amorphous solids. The Journal of Chemical Physics. 2008; **129**:144103

[37] Chu MY, Rahaman MN, De Jonghe LC, Brook RJ. Effect of heating rate on sintering and coarsening. Journal of the American Ceramic Society. 1991; **74**:1217-1225

[38] Bykov Y, Holoptsev V, Makino Y, Miyake S, Plotnikov I, Ueno T. Kinetics of densification and phase transformation at microwave sintering of silicon nitride with alumina and yttria or ytterbia as additives. Journal of the Japan Society of Powder and Powder Metallurgy. 2001;**48**(6):558-564

[39] Lange FF. Sinterability of agglomerated powders. Journal of the American Ceramic Society. 1984;**67**: 83-89

[40] German RM. Thermodynamics of sintering. In: Fang ZZ, editor. Sintering of Advanced Materials. Fundamentals and processes: Woodhead publishing; 2010

Chapter 2

# Deposition of Advanced Ceramic Coatings by Thermal Spraying

*Eugeni Cañas, Rut Benavente, Amparo Borrell and Mª Dolores Salvador*

## Abstract

Advanced ceramic coatings have been largely used in several industrial fields such as aerospace, automotive, power generation, medical or petrochemical, in order to protect or functionalise the surface of different materials. In modern industries, thermal spray processes are the most used ones to manufacture advanced ceramic coatings due to their cost advantages, flexibility and efficiency in processing ceramic materials, especially those with high melting temperature. This chapter provides a brief overview of the progress and current state of different thermal sprayed ceramics and summarises the future trend in this field. Therefore, various advanced ceramics, such as yttria-stabilised zirconia, alumina, hydroxyapatite and bioactive glasses, have been selected for analysis and discussion.

**Keywords:** advanced ceramics, ceramic coatings, thermal spraying, coating technologies, material functionallity

## 1. Introduction

Beyond polymers and metals, ceramics are one of the three major classes of materials. The word ceramic is originally from the Greek term "keramicos", that is a burnt material [1]. Nowadays, ceramics are highly employed materials, and according to their properties and applications, such materials are divided into two different categories, i.e., traditional and advanced ceramic materials [1, 2].

On the one hand, there are the traditional ceramics. These materials are entirely based on natural raw materials (clays, micas, quartz and feldspars) and are characterised by non-accurate defined properties and low reproducibility and reliability [2–5]. This category includes high-volume products such as bricks, tiles and pottery [2–5].

On the other hand, there are the advanced ceramics, that is, novel ceramic materials synthesised from chemicals of high purity, which exhibit superior properties (particle size distribution, composition, grain size, purity, etc.) tailored to one or various specific applications that require higher performance [2–6]. This category includes oxides (aluminates, titanites, zirconates, etc.), nitrides, borides and carbides [4].

Both categories also differ in the way in which they are shaped and processed, which usually involves more sophisticated steps for advanced ceramics, as well as in the way they are finished and characterised [1, 2, 4].

Nowadays, traditional ceramic materials still account for a large volume of production in the ceramics industry, but advanced ceramics' interest has exponentially grown during the last few years [5].

Advanced ceramic materials, also referred to as high-performance ceramics, engineering ceramics or technical ceramics [1, 6], are usually categorised into simple or complex classifications, based on different aspects such as their application, composition or structure [6–8]. Nevertheless, it could be interesting to divide them from an industrial point of view. According to that, advance ceramics could be grouped into five different categories, as shown in **Figure 1** [9].

Moving in a clockwise direction there are the two largest categories, that is, functional and structural ceramics. The first group includes ceramics for electrical and magnetic applications, while the second involves monoliths and composites of oxides, nitrides, borides and carbides [9]. Then, there are three specific groups, bioceramics (contains materials such as hydroxyapatite, bioactive glasses or alumina), special glasses (intended for optoelectronics applications or withstand fire among others) and ceramic coatings (advanced coatings made from most of the previous materials) [9].

Concerning ceramic coatings, these are typically manufactured through vapour deposition (both physical and chemical), sol–gel deposition and laser processes, but

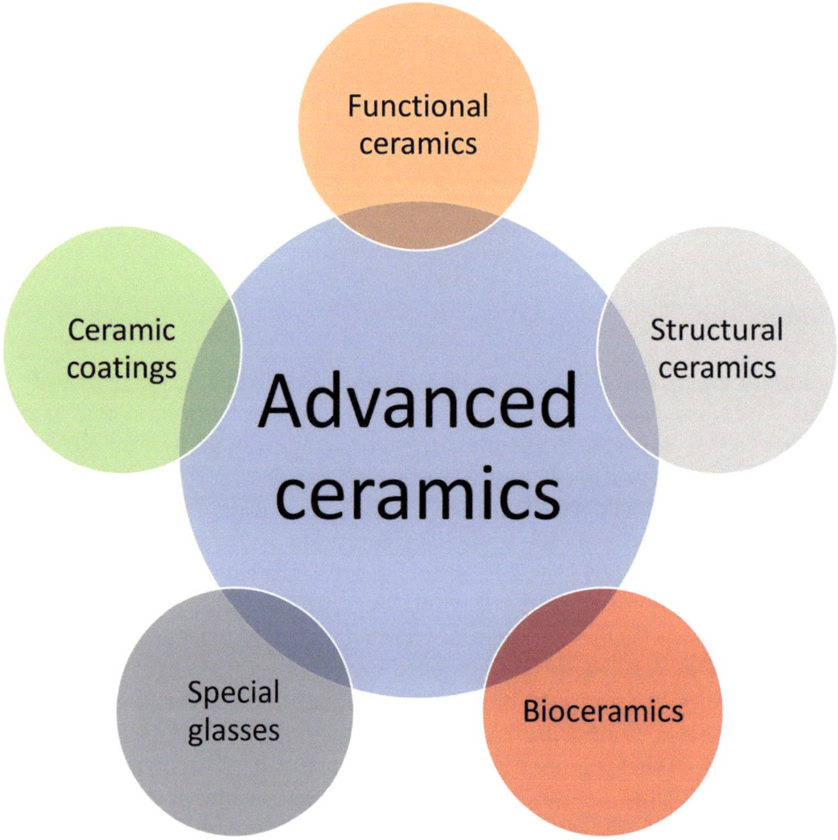


**Figure 1.**
*Possible classification of advanced ceramic materials from an industrial point of view.*

these methods result in low deposition speed coupled with high process costs [9–12]. Therefore, thermal spraying could be an alternative to these methods since it not only allows the manufacture of ceramic coatings with higher deposition speed and low process costs, but also makes it possible to obtain thicker coatings [10].

The aim of this chapter is to carry out an overview of the field of ceramic coatings' deposition by thermal spraying based on different examples of advanced ceramics.

## 2. Fundamentals of thermal spraying

The term thermal spraying encompasses a family of coating manufacturing techniques, in which metals, ceramics, polymers and mixtures of these materials are deposited onto a specific substrate in a fully or partially molten state [13, 14].

In thermal spraying, a given material (also called feedstock) is subjected to an energy source, as shown in **Figure 2** [15]. This source takes over the feedstock fragmentation into fine particles and transmission of heat and kinetic energy to them, so that the material is transformed into small droplets in a molten or semi-molten state and accelerated towards the substrate [15]. Then, upon impact on the substrate, these droplets (known as splats) suddenly cool down, flatten and become adhered to the substrate. Finally, the coating builds up by the stacking of the droplets.

In addition to the advantages described in the previous section (manufacture of thick coatings, higher deposition speed and low process costs), thermal spraying is a very versatile technology as it can be employed to deposit a large number of materials onto a wide variety of substrates. This is because this technology is capable of fluxing materials with high melting temperature without subjecting the substrate to excessive heating, since the range of temperatures to which the substrate is exposed is very wide (from 100 to 700°C approximately) [14, 16].

Due to its versatility, thermal spraying technology can be used to develop resistant coatings against corrosion, wear and high temperatures for several industrial fields such as power generation, petrochemical, automotive and aerospace among others. Moreover, improvements in both feedstocks and thermal spray processes during the last decades have allowed a significant growth of this technology in new markets, such as biomedical, dielectric and electronic coatings [14, 16].

In thermal spraying, the energy source used can be a flame from a combustion, an electric arc, a thermal plasma or kinetic energy from a decompressed gas [13–17].

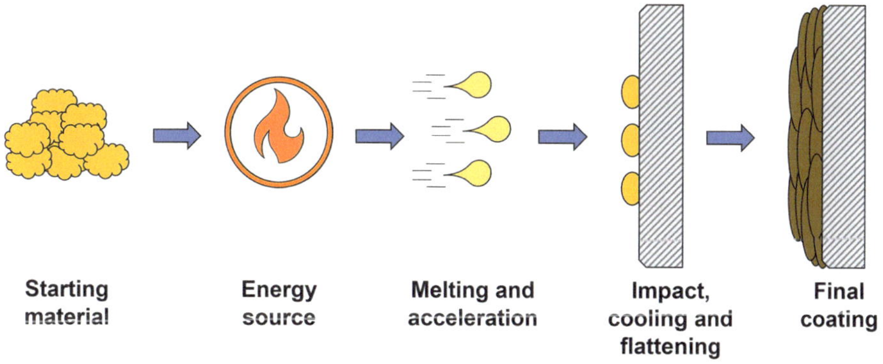


**Figure 2.**
*General process of coatings manufactured by thermal spraying. Figure adapted from [15].*

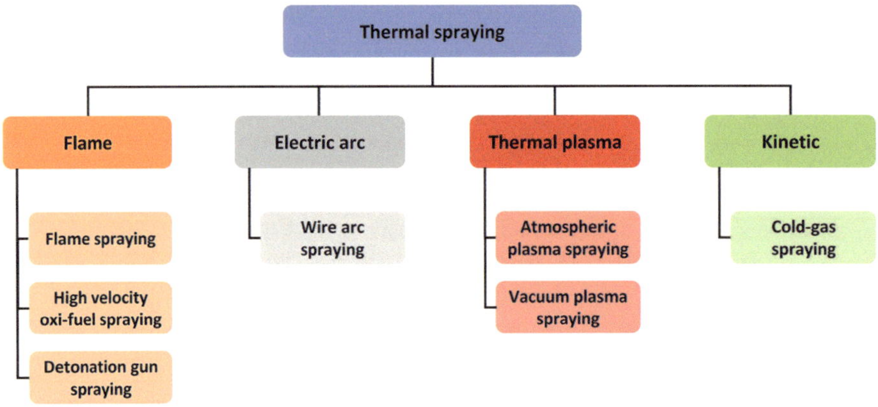


**Figure 3.**
*Thermal spraying family tree according to the energy source. Figure adapted from [14, 16].*

According to the energy source employed, all thermal spraying techniques are classified into four different families [14–17], as displayed in **Figure 3**.

Among all the techniques from the thermal spraying family, advanced ceramics are mainly deposited by plasma spraying [18]. Nevertheless, high velocity oxy-fuel spraying is also starting to receive interest due to the higher density and compactness of the coatings obtained by the high speed at which the material reaches the substrate.

In atmospheric plasma spraying (APS), the energy source is a thermal plasma, which is generated after ionisation of a gaseous mixture (typically argon, hydrogen, helium and mixtures of these) by passing through an electric arc, and recombination of the ions as they exit through a nozzle. The resulting plasma could achieve about 10,000–15,000 K in the hot spot of its core, and once generated, the feedstock is fed directly into the plasma plume where it melts and is accelerated towards the substrate [16, 17].

In high-velocity oxy-fuel spraying (HVOF), a mixture of fuel (typically different hydrocarbon gases and liquids) and oxygen is fed into a continuously pressurised combustion chamber, where it burns at high temperatures. As a result, combustion gases are generated which, due to the pressure difference, are expelled from the combustion chamber and forced through a nozzle where they reach a high velocity. Then, the feedstock is introduced into the stream of combustion gases coming from the combustion chamber, where it melts due to the high temperature of the combustion gases and is accelerated towards the substrate due to the high velocity of the combustion gases [16, 17].

Regardless of the technique, the most employed type of feedstock is powder materials, although over the last two decades the trend in thermal spraying has been towards the use of liquid materials with the aim of obtaining nanostructured coatings with improved properties. The liquid materials used are particle suspensions and solution precursors, and depending on whether one or the other is used, variants of the above techniques emerge, i.e., suspension plasma spraying (SPS), solution precursor plasma spraying (SPPS) and high velocity suspension flame spraying (HVSFS). Suspensions are usually prepared from submicron and nanoparticles using water, ethanol or other organics as solvents, whereas solution precursors are usually made by dissolving alkoxides, salts and acetates in water or ethanol, resulting in purer feedstocks as there are no contaminations from feedstock processing steps such as milling, mixing or spray-drying.

In either method, the behaviour of the material used after being subjected to the corresponding energy source is different. In the case of powders, the material only melts and is accelerated towards the substrate. When injecting the suspension, the material is first split into small suspension droplets. This splitting phenomenon is known as aerodynamic fragmentation and is due to a shear produced on the suspension by the inertia of the plasma plume. The droplet size is influenced by the viscosity of the suspension and the surface tension of the dispersant medium. Immediately after fragmentation, the dispersant starts to evaporate rapidly. From this point on, most of the fine particles contained in the suspension droplets start to sinter instantaneously forming granules, which melt and impact against the substrate to form the coating [19]. Concerning solution precursors, the processes taking place in the plasma plume are even more complex than in the case of suspensions as the material must be formed in the plasma plume itself. When the solution is injected, as in the previous case, the first stages are fragmentation and evaporation of the dispersant medium in microseconds. This is followed by precipitation of the solid in the droplet as the dispersant evaporates. For very small particles, this precipitation is complete over the entire volume of the particle, whereas, for large particles, a crust of material first precipitates on the surface of the droplet and then moves towards the centre of the droplet. Sometimes, the pressure of the liquid inside the droplet causes the initial crust to break and the process starts all over again in the resulting droplet. This is followed by pyrolysis of the precipitated solid and sintering of the formed particles into small granules, which melt and impact against the substrate [19]. The detailed processes for liquid materials correspond to the common cases, although these processes may vary depending on the particle trajectory and the processing parameters used [19].

Both techniques are complex, and the quality and final properties of the manufactured coatings depend on both the temperature and the velocity reached by the material inside the plasma plume, as well as the state of the substrate when the material impacts. At the same time, both temperature and velocity are function of more than 50 parameters. Nonetheless, some of these parameters have a more noticeable effect than the others. Basically, the main process parameters are the type of feedstock, its properties (particle size and flowability for powders and particle size, viscosity and surface tension for liquids) and how it is injected (radially or axially to the plasma plume), the flow rate and type of gases, the input energy (for plasma) and pressure of the chamber (for HVOF), the spraying distance and the state of the substrate (surface temperature and topography) [16, 17].

## 3. Manufacture of thermally sprayed ceramic coatings

### 3.1 Alumina coatings

Alumina or aluminium oxide ($Al_2O_3$) is a cheap and hard advanced ceramic, which has good wear, high hardness and corrosion, electric and thermal resistance [20]. Consequently, it is one of the most flexible and attractive advanced ceramics used in thermal spraying to be deposited onto metallic substrates for wear and erosion resistance, corrosion protection and both thermal and electrical insulation [21–23].

The manufacture of this kind of coating is commonly done by APS from powder feedstocks, although HVOF is also studied [24, 25]. In fact, with the utilisation of HVOF, the resulting coatings are less porous and compacted, and hence harder and tougher with higher electrical resistivity [20, 22, 24, 26], whereas with APS a more porous coating is obtained with better thermal insulation [26].

Regardless of the spraying technique, the main drawback of employing this oxide is that the resulting coatings are mainly composed of metastable phases, such as $\gamma$-$Al_2O_3$ and $\delta$-$Al_2O_3$, which negatively affect the long-term stability of the as-sprayed coatings and their final properties, especially in humid environments (corrosion protection in marine) and electrical insulation applications, especially APS coatings [22, 26].

Several attempts were made to increase the $\alpha$-phase present in the as-sprayed coatings, including post-heat treatments at temperatures around 1200–1300°C (unsuitable temperatures for many metallic substrates) or alloying with other oxides [27, 28]. For that purpose, chromia is the most employed oxide since an addition of 20 wt% of chromic oxide ($Cr_2O_3$) gives rise to a solid solution with $Al_2O_3$ in the $\alpha$-modification [29].

In the last decade, it has been found that with the utilisation of liquid feedstocks, instead of powders, it is possible to retain a higher amount of $\alpha$-phase (up to 70%) without the necessity of a posttreatment [30, 31]. Moreover, the electrical, mechanical properties and wear behaviour could be improved using this type of feedstock [32, 33], when compared with the coatings manufactured by conventional thermal spraying methods based on powder feedstocks.

Finally, alumina is also sprayed with zirconia to increase the toughness of the coatings and with titania to improve mechanical and tribological properties [34–37].

### 3.2 Yttria-stabilised zirconia coatings

Zirconia or zirconium oxide ($ZrO_2$) could be a suitable candidate to be used in the aerospace, automotive and power generation sectors to manufacture what is known as thermal barrier coatings (TBCs), that is, thermally sprayed ceramic coatings used to protect different metallic components of turbines and engines against corrosion and oxidation and enhance their durability and efficiency [38, 39].

This advanced ceramic has gained interest due to its excellent shock resistance, low thermal conductivity and moderate coefficaient of thermal expansion (compared to other ceramics) [39, 40].

Despite all of this, the use of pure zirconia in high temperature applications is limited due to its polymorphism, as shown in **Figure 4**. During cooling, the transformation from tetragonal phase to monoclinic phase leads to a volume reduction of approximately 4%, which results in the fracture and delamination of the coating [40, 41].

Consequently, a lattice stabilising agent is used, blocking the transformation from tetragonal to monoclinic and giving rise to a metastable tetragonal phase without volume change [41]. These agents are usually divalent, trivalent or tetravalent cations such as $Ca^{+2}$, $Y^{+3}$ or $Ce^{+4}$, which are introduced in different quantities depending on the phase to be obtained, whose purpose is to generate oxygen vacancies in the

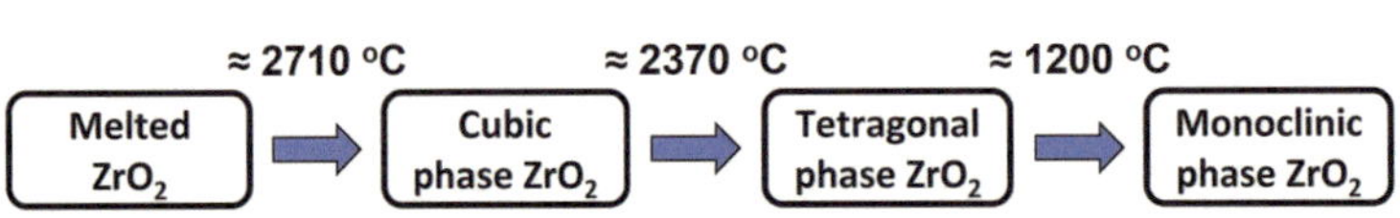


**Figure 4.**
*Phase transformations of $ZrO_2$ during cooling [40].*

crystalline network until the coordination number of Zr is reduced from 8 to around 7.5 [42]. In the case of thermal spraying, the most common agent used is yttrium oxide ($Y_2O_3$), due to the low amount needed to stabilise the compound, and its amount varies between 7 and 9 wt% (≈4 mol%) as the desired phase is the tetragonal one (which presents the lowest thermal conductivity, i.e., ≈2 W/m K) [42–45]. The resulting material is called yttria-stabilised zirconia or YSZ.

Among the thermal spraying techniques, TBCs are usually manufactured by APS from powder feedstocks [46, 47]. Employing this method, a coating is obtained with a lamellar or layered structure due to the stacking of the splats. In addition, the latter are separated by intersplat pores resulting from their rapid solidification, very fine voids formed by incomplete splat contact or presence of unmelted particles and cracks due to thermal stresses [46, 47]. Although this type of microstructure could be suitable for reducing the thermal conductivity of the coating, it negatively affects its performance, especially the thermal cycle life of the coating which is limited by insufficient strain compliance [47]. Moreover, improvement of the efficiency of gas turbines by further increases of the combustion and cooling technology in combination with higher turbine inlet temperatures implies that yttria-stabilised zirconia (4 mol% yttria) is confronted with certain limitations due to sintering and phase transformation [48]. As a result, both thermal conductivity and stiffness of the coating increase, resulting in acceleration of the spallation of the coating [48, 49].

With the aim of enhancing the TBCs' performance, the focus has been on the modification or improvement of the coating microstructure. One first approach was the deposition of coatings from submicron and nanostructured powders introduced in the plasma plume as spray-dried granules obtained from suspensions [50–52]. Coatings with a bimodal microstructure were obtained, that is, partially melted agglomerates (maintaining nanoparticles from the starting feedstock) surrounded by fully melted areas which act as matrix with superior properties compared with the conventional APS coatings from micrometric powders [52].

However, the definitive approach was the introduction of liquid feedstocks into the field of thermal spraying, and instead of spray-drying particle suspensions, these and precursor solutions began to be directly deposited by SPS and SPPS. Despite the low deposition rate, using these techniques, lower thermal conductivities were obtained compared to APS coatings [19]. In addition, by employing liquid feedstocks a totally new microstructure was obtained which is known as the "columnar structure". According to Vanevery et al. [53], when particles less than ≈3 μm in size are used, after melting and before reaching the substrate, the plasma drag forces during substrate impingement dominate the droplet inertia and redirect the droplet velocity from normal to along to the substrate surface. Consequently, droplets impact preferentially on asperities, generating deposits that grow to become columnar structures separated by linear porosity bands, which improve the performance of the coating as they stop the propagation of transversal cracks. Moreover, this effect is more marked when ethanol or organics are used as solvents, as they possess lower surface tension than water and hence, the liquid drop injected into the plasma torch will be less in size, resulting in a few number of agglomerated particles after evaporation of the solvent.

Other ways to improve the performance of the TBCs are by doping the YSZ with rare earths or by mixing YSZ and zirconates with pyrochlore structure, producing TBCs with even lower thermal conductivity and higher stability at temperature [19]. In both cases, the utilisation of SPS and SPPS facilitates this task, as both produce thinner layers than APS. Moreover, the employment of solution precursors highly facilitates the modification of the composition.

Regarding the doping of YSZ, typical dopants used are $Nd_2O_3$, $Yb_2O_3$ or $Gd_2O_3$ [19]. Nitrate precursor of the desired rare earth is commonly dissolved into the YSZ suspension or mixed with the precursors employed for YSZ. This gives a YSZ matrix with the rare earth oxides embedded inside as defects. Besides, YSZ-doped coatings can also be manufactured by hybrid plasma spraying, where a powder feedstock (YSZ) and a liquid feedstock (rare earth precursor) are deposited at the same time but are injected along different paths and at different points of the plasma plume [49].

Concerning the zirconates, the most employed ones are gadolinium zirconate ($Gd_2Zr_2O_7$) and lanthanum zirconate ($La_2Zr_2O_7$) [19]. Both materials present lower thermal conductivity, higher melting point, higher sintering resistance and higher phase stability than YSZ. Their drawback is their very low thermal expansion coefficient, preventing the use of these materials on their own [19]. Therefore, zirconates are deposited in combination with YSZ as intermediate layer using two types of layouts as shown in **Figure 5**, i.e., multi-layered coatings, in which layers of each material are deposited, and functionally graded coatings, where the composition is gradually modified from the bottom of the coating to the top of the coating.

However, functionally graded coatings present better performance compared to multi-layered, since the latter may result in failures due to thermal expansion mismatch between the layers [49].

Another application of yttria-stabilised zirconia (8 mol% yttria) is to employ it as a coating for metal-supported solid oxide fuel cells (SOFCs), more concretely covering the electrolytes [54]. Several attempts were made to deposit this type of coating by APS and high velocity oxy-fuel (HVOF). However, the utilisation of SPS and SPPS results in thinner and more uniform electrolyte coatings [55].

### 3.3 Zirconium silicate coatings

Zirconium silicate or zircon ($ZrSiO_4$) is a ceramic material which can be found naturally in zircon sands. Zircon could be a cheap and suitable candidate to replace yttria-stabilised zirconia in thermal barrier coatings as it possesses excellent thermal shock resistance, low thermal expansion coefficient and low thermal conductivity [56, 57]. Furthermore, its dielectric properties, corrosion resistance and the fact of being chemically inert at low and high temperatures allow its use in electrical applications and as an environmental barrier coating (EBC), respectively [58, 59].

Zircon coatings are typically manufactured from powder feedstocks by plasma spraying [57, 60–63]. The main problem of using this material is that zircon

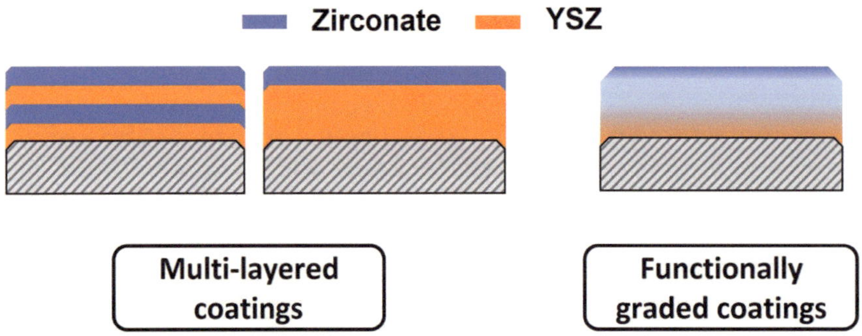


**Figure 5.**
*Layout of multi-layered and functionally graded coatings.*

decomposes in $ZrO_2$ and $SiO_2$ due to the high temperatures of the plasma plume, and the fast cooling prevents the recombination of both oxides. In addition, the different melting temperatures of $ZrO_2$ and $SiO_2$ give rise to a coating composed of crystalline $ZrO_2$ (tetragonal and monoclinic as it is not stabilised) as a major phase, a glassy solidified $SiO_2$ as a secondary phase and no zircon or only a small amount as a residual phase [56–63].

Recently, a novel approach was done based on liquid feedstocks. Aqueous suspensions of zircon particles were prepared and deposited by SPS modifying different parameters related to the technique [64]. It has been found that, due to the removal of part of the energy of the plasma plume to evaporate the water, a significant proportion of the starting zircon is preserved in the final microstructure, which remains around 20%, regardless of the processing conditions selected [64].

### 3.4 Hydroxyapatite and bioactive glass coatings

Coatings from bioceramic materials can also be manufactured by thermal spraying. Concretely, in load-bearing applications such as dental and orthopaedic metallic implants where the biomaterials cannot be used by themselves due to their brittleness. Thus, a composite is obtained that combines good mechanical properties (provided by the metallic substrate) with good osseointegration (provided by the osseoinductive ceramic layer), hence avoiding corrosion and encapsulation of the implant by fibrous tissue.

For that purpose, the most employed bioceramic is synthetic hydroxyapatite or HAp, with a stoichiometric chemical formula of $Ca_{10}(PO_4)_6(OH)_2$, a molar Ca:P ratio of 1.67 and a composition close to the mineral phase of bone [65, 66]. Coatings of HA are typically deposited by APS and HVOF from powder feedstocks [65–68]. In fact, these techniques are approved by the US Food and Drug Administration (FDA) for the deposition of HA coatings [65, 67].

Since these coatings do not require high thickness to develop its function, they are easily deposited by APS. The most important drawback of plasma-sprayed HAp coatings is to maintain the purity and crystallinity of the initial feedstock in the as-sprayed coatings. Due to the high temperatures in plasma spraying, HAp decomposes and suffers partial dehydroxylation during the manufacture of the coatings, and due to the rapid cooling it results in a deposited layer that combines crystalline HAp, amorphous phase and different calcium phosphates (tricalcium phosphate, tetracalcium phosphate and calcium oxide) [69, 70].

Both amorphous phase and calcium phosphates are undesirable, as they increase the dissolution rate and resorption of the coating, leading to a failure of the coating before it can perform its intended function and negatively affect coating's fixation to the implant and their long-term stability [69, 70]. Although several attempts were done by different researchers by modifying the most important parameters of APS, it was not possible to obtain a fully crystalline HAp coating [71, 72]. Only after a post-treatment, the full crystallinity was reached [72].

High velocity oxy-fuel (HVOF) has also been employed to deposit HAp coatings as an alternative to APS [73–76]. The lower temperatures and higher velocities of this method give rise to denser, less porous and better adhered coatings than in APS with higher crystallinity. However, it is still needed to conduct a posttreatment in order to reach fully crystalline HAp coating [76].

In both methods (APS and HVOF), it is also common to deposit HAp with other materials such as Ti, $TiO_2$ or $ZrO_2$ to improve wear resistance adhesion and fracture toughness [73, 77–80].

As in the other materials exposed in previous sections, HAp was also affected by the introduction of liquid feedstocks in the thermal spraying field, being this material deposited by means of SPS, SPPS and HVSFS [68, 81–88].

When depositing coatings by SPS, it is possible to obtain ceramic layers with a higher degree of crystallinity than the APS counterparts. This is mainly due to the trajectory of the particles in the plasma plume and the lower energy reaching them due to the evaporation of water (in the case of using this solvent) [81–83]. Regarding the trajectory, as the particles are lower in size they are more easily affected by the drag forces of the plasma plume, so that not all of them penetrate the centre of the plume, especially when the material is injected radially. As a result, a coating composed of two different zones is obtained, a dense zone formed from melted particles and a sintered zone composed of grain which were sintered instead of melting during flying across the plasma plume, preserving the crystallinity of the HAp feedstock [81, 82]. Furthermore, the utilisation of HVFSF gives rise to higher crystallinity compared with APS coatings, although the deposition efficiency is much lower [68].

Moreover, when SPPS was employed, deposition efficiency was much lower than in SPS, but the resulting coatings present higher crystallinity than those deposited by APS [86, 87]. Nevertheless, Aruna et al. have proven that SPS HAp coatings continue to be more promising than SPPS [88].

Finally, the utilisation of liquid feedstocks, especially solution precursors, greatly facilitates the modification of the composition with the aim of adding dopant ions such as $Cu^{+2}$ or $Zn^{+2}$ to improve the performance of the as-sprayed coatings [89, 90].

Bioactive glasses are the other type of bioaceramic that is under study for the deposition of bioactive coatings by means of thermal spraying. The term bioactive glass refers to a vitreous material that is biocompatible with tissues and body fluids as well as being non-carcinogenic, non-mutagenic and non-antigenic [91]. The first glass in this category is known as bioactive glass 45S5 whose composition is 45.0% $SiO_2$, 24.5% CaO, 24.5% $Na_2O$ and 6.0% $P_2O_5$ (in wt%) [91].

From this glass, several researchers have developed variants over time by slightly modifying the composition of this glass (substitution and/or elimination of oxides) depending on the application or method of synthesis. Some examples are the bioactive glass 58S [92], without sodium, as it is typically synthesised by the sol–gel method, the Bio-K [93], which includes potassium instead of sodium in order to reduce the crystallisation tendency of the glass, or the BGMS10 [94], which includes magnesium and strontium to highly increase the crystallisation temperature.

These glasses are emerging as substitutes for HAp in the manufacture of coatings by thermal spraying, since, although both materials have good biocompatibility and similar mechanical properties [95], bioactive glasses have a higher bioactivity index [95, 96], are osteoinductive, osteoconductive and osseointegrative, and during deposition by thermal spraying, they maintain their amorphous character (a requirement for their bioactivity) [65].

Like HAp, these glasses are deposited by different techniques, i.e., APS, SPS, HVSFS and SPPS [94, 97–102]. The use of liquid materials results in more porous and rougher microstructures, which increase the bioactivity of the coatings. However, there is still a lot of research ahead since, as indicated in two very recent reviews [103, 104], regardless of the spraying technique employed, the long-term stability of this kind of coating is not satisfactory yet and their adhesion to the substrate must be improved.

## 4. Conclusions and future trends in thermally sprayed ceramics

In this chapter, a brief overview of the progress and current state of different thermal sprayed ceramics has been carried out. From the previous sections, it can be seen how thermal spraying can play an important role in obtaining coatings from advanced ceramic materials. After discussing several examples of materials, it can be appreciated how it is possible to obtain different microstructures and properties depending on the spraying technique used. However, these techniques are so complex and involve so many variables. Therefore, it is very difficult to decide which technique is the definitive one for each material, as this depends on the application for which the coating is intended.

By contrast, the use of liquid feedstocks (particle suspensions and solution precursors) results in all cases in an improvement of both the microstructure and properties of the coatings. However, there is still a lot of research work to be done in order to fully understand the interaction between liquid feedstocks and the energy source used for their deposition, especially in the case of solution precursors, the effect of this interaction on the microstructure obtained and how it relates to the final properties, the aim of being able to scale the employment of these materials at industrial level.

Therefore, it could be said that the future of thermal spraying lies in the use of liquid filler materials, especially precursor solutions, since their synthesis is simpler from a processing point of view and results in materials whose composition is easier to modify and much purer.

## Acknowledgements

Authors of the present work would like to acknowledge for the grants PID2021-128548OB-C21 funded by MCIN/AEI/10.13039/501100011033 and "ERDF A way of making Europe", the Margarita Salas postdoctoral contract of Eugeni Cañas (MGS/2022/20) financed by the European Union—NextGenerationEU and Grant CIGE/2021/076 funded by Generalitat Valenciana.

## Conflict of interest

The authors declare no conflict of interest.

## Author details

Eugeni Cañas[1*], Rut Benavente[2], Amparo Borrell[2] and Mª Dolores Salvador[2]

1 Instituto Universitario de Tecnología Cerámica, Universitat Jaume I, Castellón, Spain

2 Instituto de Tecnología de Materiales, Universitat Politècnica de València, Valencia, Spain

*Address all correspondence to: eugeni.canas@itc.uji.es

## References

[1] Ayode Otitoju T, Ugochukwu Okoye P, Chen G, et al. Advanced ceramic components: Materials, fabrication, and applications. Journal of Industrial and Engineering Chemistry. 2020;**85**:34-65

[2] Carter CB, Norton MG. Ceramic Materials: Science and Engineering. New York: Springer; 2013

[3] Heimann RB. Classic and Advanced Ceramics: From Fundamentals to Applications. Weinheim: Wiley-VCH; 2010

[4] Salamon D. Advanced Ceramics. 1st ed. Waltham: Butterworth-Heinemann; 2014

[5] Rahaman MN. Ceramic Processing and Sintering. 2nd ed. Boca Ratón: CRC Press; 2010

[6] Lakhdar Y, Tuck C, Binner J, et al. Additive manufacturing of advanced ceramic materials. Progress in Materials Science. 2021;**116**:100736

[7] Kurian M, Thankachan S. Introduction: Ceramics classification and applications. In: Kurian M, Thankachan S, Nair SS, editors. Ceramic Catalysts: Materials, Synthesis and Applications. Amsterdam: Elsevier; 2023. pp. 1-17

[8] Riedel R, Chen IW, editors. Ceramics Science and Technology. Vol. 1. Structures. Weinheim: Wiley-VCH; 2008

[9] Rödel J, Kounga ABN, Weissenberger-Eibl M, et al. Development of a roadmap for advanced ceramics: 2010-2025. Journal of the European Ceramic Society. 2009;**29**:1549-1560

[10] Kassner H, Siegert R, Hathiramani D, et al. Application of suspension plasma spraying (SPS) for manufacture of ceramic coatings. Journal of Thermal Spray Technology. 2008;**17**:115-123

[11] Zanurin A, Johari NA, Alias J, et al. Research progress of sol-gel ceramic coating: A review. Materials Today Proceedings. 2022;**48**:1849-1854

[12] Horcher A, Tangermann-Gerk K, Krenkel W, et al. Advanced ceramic coatings on aluminum by laser treatment of filled organosilazane-based composites. Ceramics International. 2022;**48**:23284-23292

[13] Fauchais P, Vardelle A, Dussoubs B. Quo Vadis thermal spraying? Journal of Thermal Spray Technology. 2001;**10**:44-66

[14] Amin S, Panchal H. A Review on thermal spray coating processes. International Journal of Current Trends in Engineering & Research. 2016;**2**:556-563

[15] Davis J, editor. Handbook of Thermal Spray Technology. Ohio: ASM International; 2004

[16] Espallargas N, editor. Future Development of Thermal Spray Coatings: Types, Designs, Manufacture and Applications. Cambirdge: Woodhead Publishing; 2015

[17] Pawlowski L. The Science and Engineering of Thermal Spray Coatings. 2nd ed. Chichester: John Wiley and Sons; 2008

[18] Li CJ, Luo XT, Yao SW, et al. The bonding formation during thermal spraying of ceramic coatings: A review. Journal of Thermal Spray Technology. 2022;**31**(4):780-817

[19] Fan W, Bai Y. Review of suspension and solution precursor plasma sprayed thermal barrier coatings. Ceramics International. 2016;**42**:14299-14312

[20] Matikainen V, Niemi K, Koivuluoto H, et al. Abrasion, erosion and cavitation erosion wear properties of thermally sprayed alumina based coatings. Coatings. 2014;**4**:18-36

[21] Lampke T, Meyer D, Alisch G, et al. Alumina coatings obtained by thermal spraying and plasma anodising—A comparison. Surface & Coatings Technology. 2011;**206**:2012-2016

[22] Toma FL, Scheitz S, Berger LM, et al. Comparative study of the electrical properties and characteristics of thermally sprayed alumina and spinel coatings. Journal of Thermal Spray Technology. 2011;**20**:195-204

[23] Psyllaki PP, Jeandin M, Pantelis DI. Microstructure and wear mechanisms of thermal-sprayed alumina coatings. Materials Letters. 2001;**47**:77-82

[24] Bolelli G, Lusvarghi L, Manfredini T, et al. Comparison between plasma- and HVOF-sprayed ceramic coatings. Part I: Microstructure and mechanical properties. International Journal of Surface Science and Engineering. 2007;**1**:38-61

[25] Michalak M, Sokołowski P, Szala M, et al. Wear behavior analysis of $Al_2O_3$ coatings manufactured by APS and HVOF spraying processes using powder and suspension feedstocks. Coatings. 2021;**11**:879

[26] Shakhova I, Mironov E, Azarmi F, et al. Thermo-electrical properties of the alumina coatings deposited by different thermal spraying technologies. Ceramics International. 2017;**43**:15392-15401

[27] Krishnan R, Dash S, Sole RK, et al. Fabrication and characterisation of laser surface modified plasma sprayed alumina coatings. Surface Engineering. 2013;**18**:208-212

[28] Stahr CC, Saaro S, Berger LM, et al. Dependence of the stabilization of α-alumina on the spray process. Journal of Thermal Spray Technology. 2007;**16**:822-830

[29] Chráska P, Dubsky J, Neufuss K, et al. Alumina-base plasma-sprayed materials part I: Phase stability of alumina and alumina-chromia. Journal of Thermal Spray Technology. 1997;**6**:320-326

[30] Toma FL, Berger LM, Stahr CC, et al. Microstructures and functional properties of suspension-sprayed $Al_2O_3$ and $TiO_2$ coatings: An overview. Journal of Thermal Spray Technology. 2010;**19**:262-274

[31] Sivakumar G, Dusane RO, Joshi SV. A novel approach to process phase pure α-$Al_2O_3$ coatings by solution precursor plasma spraying. Journal of the European Ceramic Society. 2013;**33**:2823-2829

[32] Toma FL, Berger LM, Scheitz S, et al. Comparison of the microstructural characteristics and electrical properties of thermally sprayed $Al_2O_3$ coatings from aqueous suspensions and feedstock powders. Journal of Thermal Spray Technology. 2012;**21**:480-488

[33] Michalak M, Latka L, Sokolowski P, et al. Microstructural, mechanical and tribological properties of finely grained $Al_2O_3$ coatings obtained by SPS and S-HVOF methods. Surface & Coatings Technology. 2020;**404**:126463

[34] Kiilakoski J, Musalek R, Lukac F, et al. Evaluating the toughness of APS and HVOF-sprayed $Al_2O_3$-$ZrO_2$-coatings by in-situ- and macroscopic bending. Journal of the European Ceramic Society. 2018;**38**:1908-1918

[35] Carpio P, Salvador MD, Borrell A, et al. Alumina-zirconia coatings obtained by suspension plasma spraying from highly concentrated aqueous suspensions. Surface & Coatings Technology. 2016;**307**:713-719

[36] Thalib Basha GM, Srikanth A, Venkateshwarlu B. A critical review on nano structured coatings for alumina-titania ($Al_2O_3$-$TiO_2$) deposited by air plasma spraying process (APS). Materials Today Proceedings. 2020;**22**:1554-1562

[37] Jordan EH, Gell M, Sohn YH, et al. Fabrication and evaluation of plasma sprayed nanostructured alumina–titania coatings with superior properties. Materials Science and Engineering: A. 2001;**301**:80-89

[38] Ghasemi R, Vakilifard H. Plasma-sprayed nanostructured YSZ thermal barrier coatings: Thermal insulation capability and adhesion strength. Ceramics International. 2017;**43**:8556-8563

[39] Schlichting KW, Padture NP, Klemens PG. Thermal conductivity of dense and porous yttria-stabilized zirconia. Journal of Materials Science. 2001;**36**:3003-3010

[40] Tian YS, Chen CZ, Wang DY, et al. Recent developments in zirconia thermal barrier coatings. Surface Review and Letters. 2005;**12**:369-378

[41] Hannink RHJ, Kelly PM, Muddle BC. Transformation toughening in zirconia-containing ceramics. Journal of the American Ceramic Society. 2000;**83**:461-487

[42] Tsipas SA. Effect of dopants on the phase stability of zirconia based plasma sprayed thermal barrier coatings. Journal of the European Ceramic Society. 2010;**30**:61-72

[43] Moskal G. Criteria of assessment of powders provided to spray by the APS method for new and conventional layers type TBC. Archives of Materials Science and Engineering. 2009;**27**:29-36

[44] Clarke DR, Phillpot SR. Thermal barrier coating materials. Materials Today. 2005;**8**:22-29

[45] Hayashi H, Saitou T, Maruyama N, et al. Thermal expansion coefficient of yttria stabilized zirconia for various yttria contents. Solid State Ionics. 2005;**176**:613-619

[46] Chen LB. Yttria-stabilized zirconia thermal barrier coatings—A review. Surface Review and Letters. 2006;**13**:535-544

[47] Huang JB, Wang WZ, Li YJ, et al. A novel strategy to control the microstructure of plasma-sprayed YSZ thermal barrier coatings. Surface & Coatings Technology. 2020;**402**:126304

[48] Vaßen R, Jarligo MO, Steinke T, et al. Overview on advanced thermal barrier coatings. Surface & Coatings Technology. 2010;**205**:938-942

[49] Joshi SV, Sivakumar G, Raghuveer T, et al. Hybrid plasma-sprayed thermal barrier coatings using powder and solution precursor feedstock. Journal of Thermal Spray Technology. 2014;**23**:616-624

[50] Carpio P, Borrell A, Salvador MD, et al. Microstructure and mechanical properties of plasma spraying coatings from YSZ feedstocks comprising nano- and submicron-sized particles. Ceramics International. 2015;**41**:4108-4117

[51] Loghman-Estarki MR, Pourbafrany M, Shoja Razavi R, et al. Preparation of nanostructured YSZ granules by the spray drying method. Ceramics International. 2014;**40**:3721-3729

[52] Carpio P, Bannier E, Salvador MD, et al. Multilayer and particle size-graded YSZ coatings obtained by plasma spraying of micro- and nanostructured feedstocks. Journal of Thermal Spray Technology. 2014;**23**:1362-1372

[53] Vanevery K, Krane MJM, Trice RW, et al. Column formation in suspension plasma-sprayed coatings and resultant thermal properties. Journal of Thermal Spray Technology. 2011;**20**:817-828

[54] Marr M, Kuhn J, Metcalfe C, et al. Electrochemical performance of solid oxide fuel cells having electrolytes made by suspension and solution precursor plasma spraying. Journal of Power Sources. 2014;**245**:398-405

[55] Marr M, Kesler O. Permeability and microstructure of suspension plasma-sprayed YSZ electrolytes for SOFCs on various substrates. Journal of Thermal Spray Technology. 2012;**21**:1334-1346

[56] Schelz S, Enguehard F, Caron N, et al. Recombination of silica and zirconia into zircon by means of laser treatment of plasma-sprayed coatings. Journal of Materials Science. 2008;**43**:1948-1957

[57] Ramaswamy P, Seetharamu S, Varma KBR, et al. Thermal barrier coating application of zircon sand. Journal of Thermal Spray Technology. 1999;**8**:447-453

[58] Ctibor P, Pala Z, Nevrlá B, et al. Plasma-sprayed fine-grained zirconium silicate and its dielectric properties. Journal of Materials Engineering and Performance. 2017;**26**:2388-2393

[59] Liu L, Zheng W, Ma Z, et al. Study on water corrosion behavior of $ZrSiO_4$ materials. Journal of Advanced Ceramics. 2018;**7**:336-342

[60] Chráska P, Neufuss K, Herman H. Plasma spraying of zircon. Journal of Thermal Spray Technology. 1997;**6**:445-448

[61] Rudajevová A. Thermal properties of plasma-sprayed $ZrSiO_4$ material. Surface & Coatings Technology. 1994;**64**:47-51

[62] Suzuki M, Sodeoka S, Inoue T. Zircon-based ceramics composite coating for environmental barrier coating. Journal of Thermal Spray Technology. 2008;**17**:404-409

[63] Suzuki M, Sodeoka S, Inoue T. Structure control of plasma sprayed zircon coating by substrate preheating and post heat treatment. Materials Transactions. 2005;**46**:669-674

[64] Cañas E, Rosado E, Alcázar C, et al. Challenging zircon coatings by suspension plasma spraying. Journal of the European Ceramic Society. 2022;**42**:4369-4376

[65] Juhasz JA, Best SM. Bioactive ceramics: Processing, structures and properties. Journal of Materials Science. 2011;**47**(2):610-624

[66] Berndt C, Fahad Hasan M, Tietz U, et al. A review of hydroxyapatite coatings manufactured by thermal spray. In: Ben-Nissan B, editor. Advances in Calcium Phosphate Biomaterials. Berlin, Heidelberg: Springer; 2014. pp. 267-329

[67] Balani K, Chen Y, Harimkar SP, et al. Tribological behavior of plasma-sprayed carbon nanotube-reinforced hydroxyapatite coating in physiological solution. Acta Biomaterialia. 2007;**3**:944-951

[68] Gadow R, Killinger A, Stiegler N. Hydroxyapatite coatings for biomedical applications deposited by different thermal spray techniques. Surface & Coatings Technology. 2010;**205**:1157-1164

[69] Sun L, Berndt CC, Gross KA, et al. Material fundamentals and clinical

performance of plasma-sprayed hydroxyapatite coatings: A review. Journal of Biomedical Materials Research. 2001;**58**:570-592

[70] Cheang P, Khor KA. Addressing processing problems associated with plasma spraying of hydroxyapatite coatings. Biomaterials. 1996;**17**:537-544

[71] Sun L, Berndt CC, Grey CP. Phase, structural and microstructural investigations of plasma sprayed hydroxyapatite coatings. Materials Science and Engineering: A. 2003;**360**:70-84

[72] Xue W, Tao S, Liu X, et al. In vivo evaluation of plasma sprayed hydroxyapatite coatings having different crystallinity. Biomaterials. 2004;**25**:415-421

[73] Henao J, Cruz-bautista M, Hincapie-Bedoya J, et al. HVOF hydroxyapatite/titania-graded coatings: Microstructural, mechanical, and in vitro characterization. Journal of Thermal Spray Technology. 2018;**27**:1302-1321

[74] Lima RS, Khor KA, Li H, et al. HVOF spraying of nanostructured hydroxyapatite for biomedical applications. Materials Science and Engineering: A. 2005;**396**:181-187

[75] Khor KA, Li H, Cheang P. Significance of melt-fraction in HVOF sprayed hydroxyapatite particles, splats and coatings. Biomaterials. 2004;**25**:1177-1186

[76] Fernández J, Gaona M, Guilemany JM. Effect of heat treatments on HVOF hydroxyapatite coatings. Journal of Thermal Spray Technology. 2007;**16**:220-228

[77] Zheng X, Huang M, Ding C. Bond strength of plasma-sprayed hydroxyapatite/Ti composite coatings. Biomaterials. 2000;**21**:841-849

[78] Ong JL, Carnes DL, Bessho K. Evaluation of titanium plasma-sprayed and plasma-sprayed hydroxyapatite implants in vivo. Biomaterials. 2004;**25**:4601-4606

[79] Li H, Khor KA, Cheang P. Titanium dioxide reinforced hydroxyapatite coatings deposited by high velocity oxy-fuel (HVOF) spray. Biomaterials. 2002;**23**:85-91

[80] Li H, Khor KA, Kumar R, et al. Characterization of hydroxyapatite/nano-zirconia composite coatings deposited by high velocity oxy-fuel (HVOF) spray process. Surface & Coatings Technology. 2004;**182**:227-236

[81] Kozerski S, Pawlowski L, Jaworski R, et al. Two zones microstructure of suspension plasma sprayed hydroxyapatite coatings. Surface & Coatings Technology. 2010;**204**:1380-1387

[82] Łatka L, Pawlowski L, Chicot D, et al. Mechanical properties of suspension plasma sprayed hydroxyapatite coatings submitted to simulated body fluid. Surface & Coatings Technology. 2010;**205**:954-960

[83] Zheng B, Luo Y, Liao H, et al. Investigation of the crystallinity of suspension plasma sprayed hydroxyapatite coatings. Journal of the European Ceramic Society. 2017;**37**:5017-5021

[84] Bolelli G, Bellucci D, Cannillo V, et al. Suspension thermal spraying of hydroxyapatite: Microstructure and in vitro behaviour. Materials Science and Engineering: C. 2014;**34**:287-303

[85] Krieg P, Killinger A, Gadow R, et al. High velocity suspension flame spraying (HVSFS) of metal doped bioceramic coatings. Bioactive Materials. 2017;**2**:169

[86] Candidato RT, Sokołowski P, Pawłowski L, et al. Development of hydroxyapatite coatings by solution precursor plasma spray process and their microstructural characterization. Surface & Coatings Technology. 2017;**318**:39-49

[87] Huang Y, Song L, Huang T, et al. Characterization and formation mechanism of nano-structured hydroxyapatite coatings deposited by the liquid precursor plasma spraying process. Biomedical Materials. 2010;**5**:054113

[88] Aruna ST, Kulkarni S, Chakraborty M, et al. A comparative study on the synthesis and properties of suspension and solution precursor plasma sprayed hydroxyapatite coatings. Ceramics International. 2017;**43**:9715-9722

[89] Unabia RB, Bonebeau S, Candidato RT, et al. Investigation on the structural and microstructural properties of copper-doped hydroxyapatite coatings deposited using solution precursor plasma spraying. Journal of the European Ceramic Society. 2019;**39**:4255-4263

[90] Candidato RT, Sergi R, Jouin J, et al. Advanced microstructural study of solution precursor plasma sprayed Zn doped hydroxyapatite coatings. Journal of the European Ceramic Society. 2018;**38**:2134-2144

[91] Gurbinder K. Bioactive Glasses–Potential Biomaterials for Future Therapy. 1st ed. Switzerland: Springer Nature; 2017

[92] Jones JR. Review of bioactive glass: From Hench to hybrids. Acta Biomaterialia. 2013;**9**:4457-4486

[93] Cannillo V, Sola A. Potassium-based composition for a bioactive glass. Ceramics International. 2009;**35**:3389-3393

[94] Cañas E, Orts MJ, Sánchez E, et al. Deposition of bioactive glass coatings based on a novel composition containing strontium and magnesium. Journal of the European Ceramic Society. 2022;**42**:6213-6221

[95] Goller G, Demirkiran H, Oktar FN, et al. Processing and characterization of bioglass reinforced hydroxyapatite composites. Ceramics International. 2003;**29**:721-724

[96] Lefebvre L, Chevalier J, Gremillard L, et al. Structural transformations of bioactive glass 45S5 with thermal treatments. Acta Materialia. 2007;**55**:3305-3313

[97] Cannillo V, Sola A. Different approaches to produce coatings with bioactive glasses: Enamelling vs plasma spraying. Journal of the European Ceramic Society. 2010;**30**:2031-2039

[98] Calvo VL, Cabedo MV, Bannier E, et al. 45S5 bioactive glass coatings by atmospheric plasma spraying obtained from feedstocks prepared by different routes. Journal of Materials Science. 2014;**49**:7933-7942

[99] Cattini A, Łatka L, Bellucci D, et al. Suspension plasma sprayed bioactive glass coatings: Effects of processing on microstructure, mechanical properties and in-vitro behaviour. Surface & Coatings Technology. 2013;**220**:52-59

[100] Cañas E, Vicent M, Orts MJ, et al. Bioactive glass coatings by suspension plasma spraying from glycolether-based solvent feedstock. Surface & Coatings Technology. 2017;**318**:190-197

[101] Altomare L, Bellucci D, Bolelli G, et al. Microstructure and in vitro behaviour of 45S5 bioglass coatings deposited by high velocity suspension flame spraying (HVSFS). Journal of Materials Science: Materials in Medicine. 2011;**22**:1303-1319

[102] Cañas E, Orts MJ, Boccaccini AR, et al. Microstructural and in vitro characterization of 45S5 bioactive glass coatings deposited by solution precursor plasma spraying (SPPS). Surface & Coatings Technology. 2019;**371**:151-160

[103] Sergi R, Bellucci D, Cannillo V. A comprehensive review of bioactive glass coatings: State of the art, challenges and future perspectives. Coatings. 2020;**10**:757

[104] Garrido B, Dosta S, Cano IG. Bioactive glass coatings obtained by thermal spray: Current status and future challenges. Boletín de la Sociedad Española de Cerámica y Vidrio. 2022;**61**:516-530

Chapter 3

# Study of SPS Sintering of Strontium-Doped Lanthanum Manganite (LSM) by Surface Modification of Powders Using DCSBD and ALD

*Amparo Borrell, Rut Benavente, René M. Guillén, María D. Salvador, Vaclav Pouchly, Martina Ilcikova, Richard Krumpolec and Rodrigo Moreno*

## Abstract

Throughout the ceramic processing cycle, it is well known that a small change in the surface energy of as-received powders can have a considerable effect on the final properties of consolidated materials. The main objective of this chapter is to describe the design and manufacture of new ceramic materials based on strontium-doped lanthanum manganites, LSM ($La_{0.8}Sr_{0.2}MnO_3$) and LSM-8YTZP composites, for cathode in solid oxide fuel cells (SOFC) applications due to their excellent properties, by modifying the surface energy of the starting powder using techniques, such as Diffuse Coplanar Surface Barrier Discharge (DCSBD) and atomic layer deposition (ALD). Subsequently, in order to evaluate the activation energy and optimise the sintering behaviour of these powders, the Spark Plasma Sintering (SPS) technique will be used. SPS allows the complete densification of pieces by fast and low-energy consumption processing.

**Keywords:** strontium-doped lanthanum manganites, spark plasma sintering, diffuse coplanar surface barrier discharge, atomic layer deposition, ceramics

## 1. Introduction

Strontium-doped lanthanum manganite ($La_{0.8}Sr_{0.2}MnO_3$) materials have a crystal structure of the perovskite-type A-B-$O_3$, where A and B represent the 12- and 6-coordinated metals, respectively. The $La_{1-x}Sr_xMnO_3$ (LSM) has attracted attention in recent years due to its excellent thermal, electronic and magnetic properties [1] with potential applications in magnetic sensors, readout heads for magnetic memories and as a cathode in solid oxide fuel cells (SOFC) [2].

One of the most important applications of these materials is solid oxide fuel cells. The electrolyte is a ceramic oxide that is generally located in the centre of the cell (although there are exceptions such as in porous metal-supported cells) to facilitate the generation of oxygen vacancies and to transport the ionic charge between the cathode and the anode (**Figure 1**) [3]. The cathode and anode are electrodes where different reactions take place: oxygen is reduced to oxide ions by consuming two electrons at the cathode, and fuel is reduced by forming two electrons at the anode. The electrodes must be porous and facilitate the transport of reactants and products through the different components [4, 5].

The cathode is one of the essential components of a SOFC. The cathode materials must meet several requirements in the operating temperature of 800–1000°C: high electronic conductivity, thermal and chemical stability, catalytic activity for oxygen reduction, and a thermal expansion and coefficients of the same order as the other SOFC components [6, 7].

The use of yttria-stabilised zirconia (Y-TZP) as an electrolyte material is customarily found in the literature, being the main material used for the fabrication of solid electrochemical devices. Its use is due to the fact that this material has high conductivity and ionic stability with oxygen. It is important to emphasise that in solid oxide fuel cells with zirconia as a base material, lanthanum manganites doped with strontium are a key point for better cell efficiency. LSM materials are commonly used as cathode material due to their high electronic conductivity and fast oxygen transfer at gas–solid interfaces, combined with their thermal expansion capability and chemical compatibility with yttria-stabilised zirconia solid electrolyte [8]. The content of yttrium oxide ($Y_2O_3$) as a dopant for zirconia is a very important factor. In terms of conductivity, zirconia doped with 8–10 mol% yttria has high ionic conductivity. Moreover, in terms of structural and mechanical properties, zirconia doped with 8 mol% yttria has high flexural and impact strength [9]. Therefore, in this chapter, LSM-8YTZP composites will be investigated due to their good mechanical properties [10].

This chapter aims to study how the heating rate, the final temperature and the dwell time affect the final properties of the LSM material and the LSM-8YTZP composite densified using the Spark Plasma Sintering technology. In a second stage, a

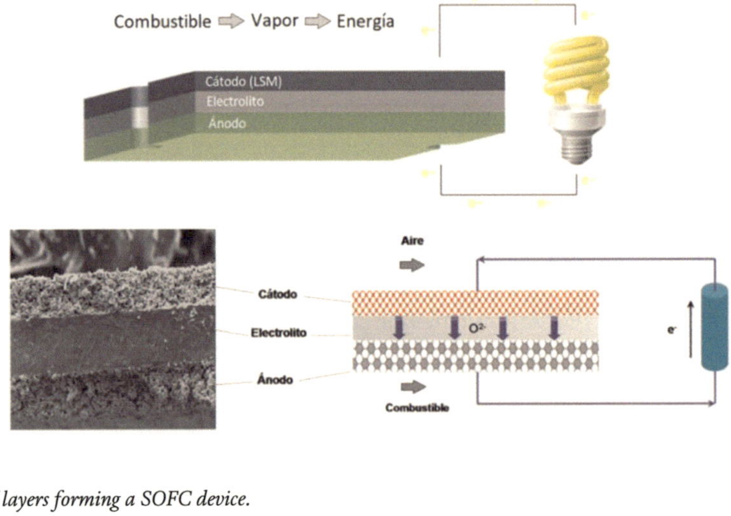


**Figure 1.**
*Diagram of layers forming a SOFC device.*

surface modification of the starting powders, both LSM and LSM-8YTZP, will be carried out using Diffuse Coplanar Surface Barrier Discharge (DCSBD) and Atomic Layer Deposition (ALD) techniques in order to identify changes in the material that may influence the sintering and, therefore, the final properties.

## 2. Materials and methods

### 2.1 Raw materials

As raw materials, commercial $La_{1-x}Sr_XMnO_3$ (x = 0.2) powder (INFRAMAT, Advanced Materials, USA) with an average particle size of $d_{50}$ = 0.25 μm, a BET-specific surface area of 2.8 $m^2/g$ and high purity commercial yttria-stabilised zirconia (YSZ) with 8 mol% $Y_2O_3$ (TZ8YS, Tosoh Corp., Japan), with an average particle size of 0.4 μm, and a BET specific surface area of 4.7 $m^2/g$ were used.

The mixture preparation was carried out after optimising the zeta potential of the LSM and 8YTZP suspensions and the rheological behaviour of the LSM-8YTZP mixture. LSM and LSM-8YTZP were prepared as concentrated suspensions in water with a solid loading of 30 vol%. The mixtures were prepared with a volume relative content of 50:50 according to a sequential addition, by first adding the deflocculant required to disperse the 8YTZP (0.5 wt.% on a dry solid basis) and then the 8YTZP powder, followed by the deflocculant required for the LSM (1.5 wt.% on a dry solid basis) before the LSM powder. The optimised slurry of the mixture was then frozen using a rotavapor (IKA, Germany) in a liquid-$N_2$ bath and subsequently freeze-dried (Cryodos-50, Telstar, Spain) at −50°C and 0.3 mPa vacuum pressure for 24 h to sublimate the ice.

The freeze-dried powders were sintered by Spark Plasma Sintering (SPS, DR. SINTER SPS-625, FUJI, Japan), and the surface energy of this powder was also modified using low-energy techniques such as Diffuse Coplanar Surface Barrier Discharge (DCSBD) and Atomic Layer Deposition (ALD).

Spark Plasma Sintering was carried out under vacuum (5–9 Pa) using a graphite mould with a diameter of 12 mm. The temperature was measured using an optical pyrometer focused on the die wall, a constant uniaxial pressure of 15 MPa was applied above 600°C, and a dwell time of 2 min was used.

### 2.2 Diffuse coplanar surface barrier discharge (DCSBD)

Diffuse Coplanar Surface Barrier Discharge (DCSBD) is a novel type of atmospheric-pressure plasma source developed for high-speed, large-area surface plasma treatments [11]. The plasma of DCSBD discharge is generated in a sub-millimetre thin layer above the dielectric plate. With a gradual increase of power input, the DCSBD gets visually more and more diffuse and homogeneous, though it is still composed of a high number of individual microdischarges [12–14]. **Figure 2** shows a scheme of the DCSBD electrode system arrangement and a general view of the DCSBD apparatus.

The number of DCSBD microdischarges generated during one discharge period is very high (in the order of tens to hundreds of single microdischarges), and the generated microdischarges are distributed above a large area of approx. 20 × 8 $cm^2$. The typical dimensions of a single microdischarge were of the order of 1 mm in length and 100 μm in diameter in the time scale of several tens of nanoseconds. The system

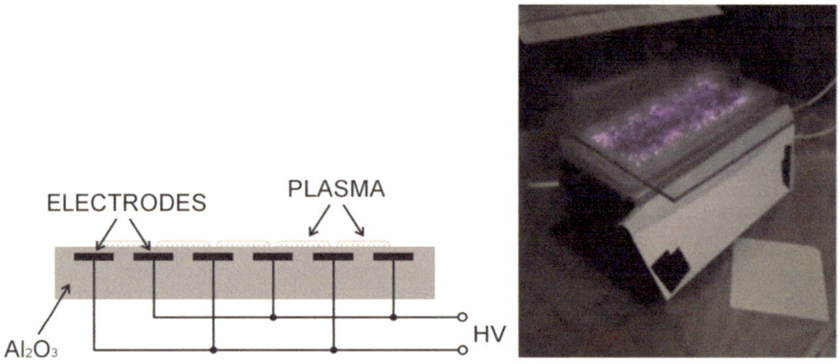


**Figure 2.**
*Cross-sectional schematics of DCSBD electrode system arrangement [15] and visual appearance of material with DCSBD apparatus.*

was powered by a 14 kHz sinusoidal high voltage of up to 20 kV peak-to-peak amplitude and an input power of 400 W, supplied by a power generator. The powdered material was activated for 45 s and removed for further ceramic processing.

### 2.3 Atomic layer deposition (ALD)

Atomic layer deposition (ALD) is a method for depositing ultra-thin films with a wide range of applications due to its unique capabilities. This procedure can increase the efficiency of electrical devices by allowing the homogeneous deposition of conformal films with controlled thickness, even on complicated three-dimensional surfaces [16].

The powder was coated in a quartz tube of ID = 20 mm and length 100 mm on both sides, partially sealed by quartz wool. This tube was placed in the centre of a cylindrical deposition chamber diameter of 50 mm and a length of 450 mm. The deposition was done at a chamber pressure of 180 Pa. Pure nitrogen (99.999%) was used as a carrier and purging gas at a total flow rate of 90 sccm (standard cubic centimetres per minute). The zinc oxide coatings were deposited using Diethylzinc (DEtZn, STREM Chemicals) as a metal source and deionised water (Merck Millipore Q-water, 18.2 MΩ) as an oxygen source (**Figure 3**).

Each deposition consisted of 20 ALD cycles starting with $H_2O$ pulse. One ZnO deposition cycle was defined by the following sequence: $H_2O$ pulse (10 s) → $N_2$ purge (30 s) → DEtZn pulse (10 s) → $N_2$ purge (30 s). Quite long pulsing and purging times were used due to the deposition of thin coatings on powders. Both DEtZn and $H_2O$

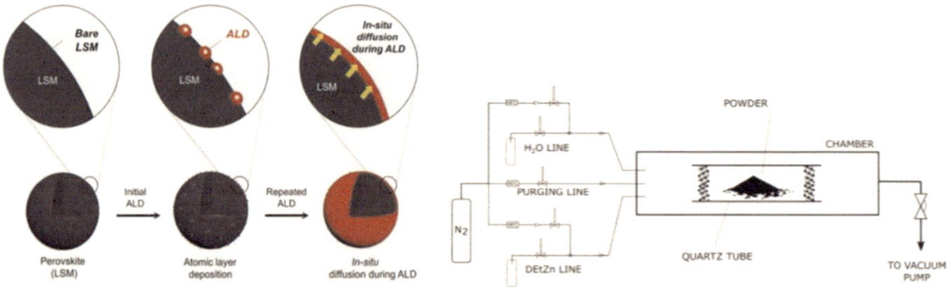


**Figure 3.**
*Experimental setup of the ALD coating equipment.*

precursors were kept at room temperature. Delivery lines were heated at 60/100°C, and the temperature in the deposition chamber was 140°C. The tube with a coated powder was removed from the deposition chamber always after cooling down to 60°C.

### 2.4 Characterisation methods

The relative densities of specimens were determined by the Archimedes' method following the ASTM-C-373 standard [17]. Theoretical density values of 6.57 and 6.41 g/cm$^3$ were used for LSM and LSM-8YTZP, respectively, to estimate the relative density of the sintered specimens.

The microstructure and grain size were analysed using field emission scanning electron microscopy (FE-SEM, S4800 Hitachi, Japan). The grain size was measured using the line interception method following the ASTM E112 standard [18], with 10 interception lines being traced in each sample using Image software.

The evaluation of mechanical characteristics (hardness and Young's modulus) was done according to the nanoindentation technique. The method used is a nanoindenter (G-200; Agilent Technologies). Tests were carried out with a Berkovich tip calibrated with silica standard and operated at a maximum depth of 2000 nm. The continuous stiffness measurement was used to determine the contact stiffness and calculate the hardness profiles and elastic modulus [19]. A matrix with 25 indentations was made for every material.

## 3. Results and discussion

### 3.1 Study of the LSM powder

#### *3.1.1 LSM sintered by spark plasma sintering*

First, a Spark Plasma Sintering (SPS) study of the influence of heating rate on LSM densification was carried out. For this purpose, three heating rates were defined, i.e., 50, 100, and 200°C/min, at a final temperature of 1200°C with a dwell time of 2 min, and a constant uniaxial pressure of 15 MPa was applied above 600°C.

In **Table 1**, the values of average grain size and relative density corresponding to the LSM material sintered by SPS at 1200°C—2 min using different heating rates can be observed.

Small changes are observed with respect to relative density and grain size values. The density value above 90% is reached with a heating rate of 200°C/min. The average grain size is around 1.80 μm, approximately, as can be seen from the microstructure in **Figure 4**. This figure shows the FESEM microstructures, at different

| Material | Sintering parameters | Heating rate (°C/min) | Grain size (μm) | Relative density (%) |
|---|---|---|---|---|
| LSM | 1200°C—2 min | 50 | 1.75 ± 0.04 | 88.0 ± 0.5 |
| | | 100 | 1.77 ± 0.05 | 88.9 ± 0.5 |
| | | 200 | 1.85 ± 0.07 | 90.2 ± 0.5 |

**Table 1.**
*Grain size and relative density values of LSM material sintered by SPS.*

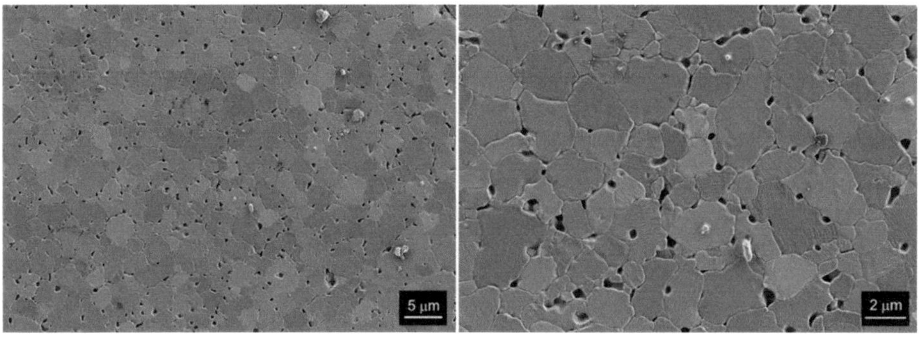


**Figure 4.**
*Microstructure of LSM material sintered by SPS at 1200°C—2 min, heating ramp 200°C/min.*

| Material | Sintering parameters | Heating rate (°C/min) | Hardness (GPa) | Young modulus (GPa) |
|---|---|---|---|---|
| LSM | 1200°C—2 min | 50 | 7.0 ± 0.3 | 108 ± 0.2 |
| | | 100 | 7.1 ± 0.3 | 109 ± 0.4 |
| | | 200 | 7.5 ± 0.2 | 120 ± 0.2 |

**Table 2.**
*Hardness and young modulus values of LSM material sintered by SPS.*

magnifications, of polished LSM material sintered by SPS at 1200°C with a heating rate of 200°C/min. It can be observed that large and small grains coexist. In general, all heating rates follow the same tendency, with slightly larger average grain sizes being obtained at higher heating rates, at which higher densification is achieved.

The values of hardness and Young's modulus can be seen in **Table 2**. The Young's modulus values of the material sintered at 50 and 100°C/min show similar values, but a slight increase is observed for the heating rate of 200°C/min. For hardness, an 8% increase is observed when the heating rate is increased from 50 to 200°C/min. This increase is very remarkable, as higher mechanical properties have been achieved at very high heating rates, which influences the total processing cycle of the material. Using a heating rate of 50°C/min, the complete sintering cycle reaches 16 min, and using a heating rate of 200°C/min, it only requires 8 min to complete the sintering cycle. The energy and time savings are therefore doubled.

### *3.1.2 LSM powders modified by diffuse coplanar surface barrier discharge (DCSBD) and sintered by SPS*

The LSM starting powders were modified by the DCSBD technique, using the parameters mentioned in Section 2.2, and then sintered by SPS at 1200°C—2 min, using the same parameters as in the previous section.

In **Table 3**, the values of average grain size and relative density corresponding to the LSM powders modified by DCSBD and sintered by SPS at 1200°C—2 min using different heating rates can be observed.

By means of the DCSBD treatment, it can be identified that the density values of the material change significantly as a function of the heating rate. Density values above 95% are reached, indicating that with this treatment, the density of the final material increases significantly. These density values will have an impact on both the

| Material DCSBD | Sintering parameters | Heating rate (°C/min) | Grain size (μm) | Relative density (%) |
|---|---|---|---|---|
| LSM | 1200°C—2 min | 50 | 1.78 ± 0.04 | 92.7 ± 0.5 |
| | | 100 | 1.80 ± 0.03 | 93.1 ± 0.5 |
| | | 200 | 1.75 ± 0.05 | 95.1 ± 0.5 |

**Table 3.**
*Grain size and relative density values of LSM powders modified by DCSBD and sintered by SPS.*

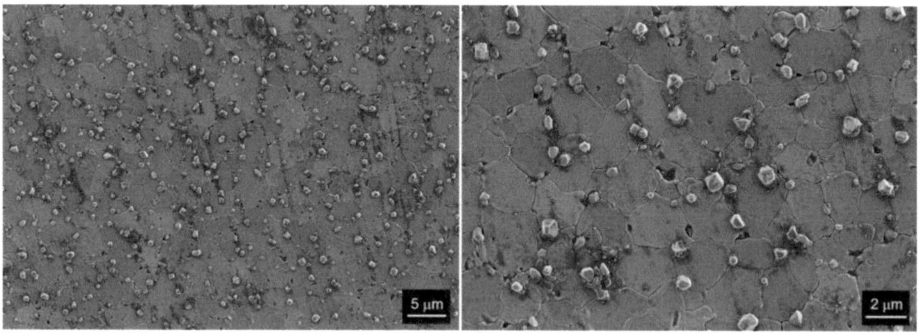


**Figure 5.**
*Microstructure of LSM material with DCSBD treatment and SPS at 1200°C—2 min, heating ramp 200°C/min.*

microstructure and the final mechanical properties. On the other hand, taking into account the grain size values in **Table 1**, it is observed that the materials without any treatment and with DCSBD treatment present similar values. This means that the DCSBD treatment enhances densification while maintaining the same grain sizes.

The FESEM microstructures of the DCSBD-treated sample sintered at 1200°C with a heating rate of 200°C/min, which has a value of 1.75 μm, can be seen in **Figure 5** at different magnifications. Smaller particles of LSM material are observed homogeneously distributed over the entire surface. This may be due to the technique used, as the number of DCSBD microdischarges generated during a discharge period is very high (in the order of tens to hundreds of individual microdischarges), and these generated microdischarges are distributed over a large surface. This surface modification positively influences the densification of the material.

**Figure 5** shows a small residual porosity, as the small LSM particles created are located at the grain boundary and help, on the one hand, to close the porosity and, on the other hand, have a tendency to inhibit grain growth.

The values of hardness and Young's modulus can be seen in **Table 4**. The mechanical properties are also influenced by the surface treatment of the powder

| Material DCSBD | Sintering parameters | Heating rate (°C/min) | Hardness (GPa) | Young modulus (GPa) |
|---|---|---|---|---|
| LSM | 1200°C—2 min | 50 | 7.7 ± 0.2 | 124 ± 0.2 |
| | | 100 | 7.9 ± 0.2 | 119 ± 0.4 |
| | | 200 | 8.3 ± 0.1 | 136 ± 0.3 |

**Table 4.**
*Hardness and young modulus values of LSM powders modified by DCSBD and sintered by SPS.*

before sintering by SPS. Using a heating rate of 200°C/min, hardness values of 8.3 GPa are achieved. Compared to the hardness value achieved with a heating rate of 50°C/min, the improvement is 8%. These hardness values are higher than those shown in **Table 2**, corresponding to the non-treated LSM material. Comparing the LSM material sintered at 1200°C with a heating ramp of 200°C/min, without treatment and the one treated with DCSBD, the difference in hardness values is about 13%. The Young's modulus values achieved using a heating ramp of 200°C/min are around 136 GPa, which represent the highest values obtained. This is in good agreement with the higher densification level.

### *3.1.3 LSM powders modified by atomic layer deposition (ALD) and sintered by SPS*

The LSM starting powders were modified by the ALD technique, using the parameters mentioned in Section 2.3, and then sintered by SPS at 1200°C—2 min, using the same parameters as in the previous section. In **Table 5**, the values of average grain size and relative density corresponding to the LSM powders modified by ALD and sintered by SPS at 1200°C—2 min using different heating rates can be observed.

The final density of the ALD surface-modified LSM sintered using different heating rates is very similar to the non-treated LSM material (**Table 1**). The FESEM microstructures of the ALD-treated sample sintered at 1200°C with a heating rate of 200°C/min, can be seen in **Figure 6**, at different magnifications.

As observed in the microphotographs, a great difference can be seen in the final grain size, which is much higher in these materials treated with the ALD technique. Porosity is observed at the grain boundaries as well as residual porosity in the interior of the grains. The average grain size in all samples is around 2.60 μm. The ZnO layer deposited by this methodology reactivates the surface of the LSM powders, leading to

| Material ALD | Sintering parameters | Heating rate (°C/min) | Grain size (μm) | Relative density (%) |
|---|---|---|---|---|
| LSM | 1200°C—2 min | 50 | 2.60 ± 0.03 | 87.3 ± 0.5 |
| | | 100 | 2.57 ± 0.05 | 88.2 ± 0.5 |
| | | 200 | 2.62 ± 0.05 | 90.5 ± 0.5 |

**Table 5.**
*Grain size and relative density values of LSM powders modified by ALD and sintered by SPS.*

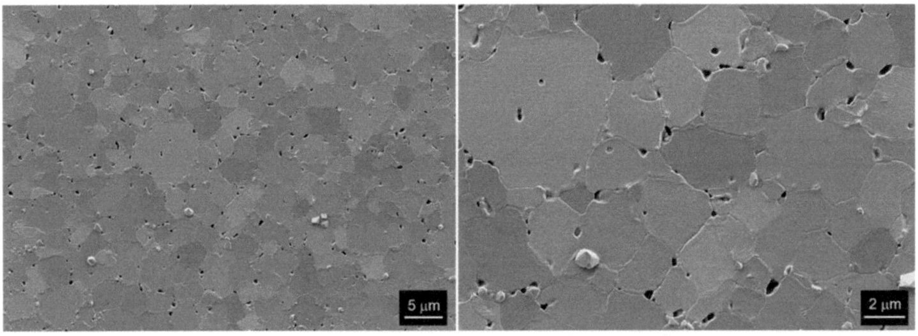


**Figure 6.**
*Microstructure of LSM material with ALD treatment and SPS at 1200°C—2 min, heating ramp 200°C/min.*

| Material ALD | Sintering parameters | Heating rate (°C/min) | Hardness (GPa) | Young modulus (GPa) |
|---|---|---|---|---|
| LSM | 1200°C—2 min | 50 | 6.1 ± 0.2 | 92 ± 0.3 |
| | | 100 | 5.9 ± 0.3 | 109 ± 0.3 |
| | | 200 | 6.4 ± 0.1 | 120 ± 0.2 |

**Table 6.**
*Hardness and young modulus values of LSM powders modified by ALD and sintered by SPS.*

an exaggerated grain growth behaviour in the sintering phase. This grain growth does not help the densification of the material; therefore, the final porosity is around 10%. This will have a negative impact on the mechanical properties, as shown below.

The values of hardness and Young's modulus can be seen in **Table 6**. It is observed, as with the non-treated LSM and the LSM treated with DCSBD, that as the heating ramp values are higher, the mechanical values also show an increase.

The mechanical values indicate that the LSM materials treated with the ALD technique do not lead to an improvement. Comparing the hardness values with those obtained for the non-treated LSM materials (**Table 2**) and those treated with DCSBD (**Table 4**), it can be seen that they are below 7.0 GPa.

If a comparison is carried out between the non-treated LSM materials and those treated with the different DCSBD and ALD techniques, it can be concluded that the treatment that provides the best results is the DCSBD technique. This technique provides lower residual porosity in the material and helps to inhibit grain growth, which leads to superior mechanical properties. In the following section, the influence of surface powder treatment techniques, DCSBD and ALD, on LSM-8YTZP composites will be analysed.

## 3.2 Study of the LSM-8YTZP powder

### *3.2.1 LSM-8YTZP sintered by spark plasma sintering*

Spark Plasma Sintering (SPS) study of the influence of heating ramps on the LSM-8YTZP material was carried out. For this purpose, the same three heating ramps were defined, i.e., 50, 100, and 200°C/min, at a final temperature of 1200°C with a dwell time of 2 min, applying a constant uniaxial pressure of 15 MPa above 600°C. The obtained results for average grain size and relative density are shown in **Table 7**.

It is observed that as the heating ramp is faster, higher relative density values are achieved. This indicates that rapid heating promotes the densification of these composites. As for the grain size values, all the composites sintered at different heating rates are around 0.36 μm. If these values are compared with those obtained in **Table 1**

| Material | Sintering parameters | Heating rate (°C/min) | Grain size (μm) | Relative density (%) |
|---|---|---|---|---|
| LSM-8YTZP | 1200°C—2 min | 50 | 0.30 ± 0.03 | 92.6 ± 0.5 |
| | | 100 | 0.36 ± 0.05 | 93.1 ± 0.5 |
| | | 200 | 0.40 ± 0.02 | 93.9 ± 0.5 |

**Table 7.**
*Grain size and relative density values of LSM-8YTZP composites sintered by SPS.*

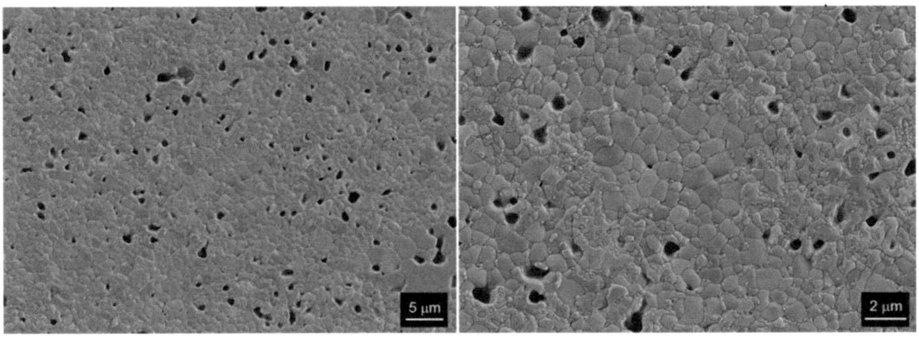


**Figure 7.**
*Microstructure of LSM-8YTZP composite sintered by SPS at 1200°C—2 min, heating ramp 200°C/min.*

| Material | Sintering parameters | Heating rate (°C/min) | Hardness (GPa) | Young modulus (GPa) |
|---|---|---|---|---|
| LSM-8YTZP | 1200°C—2 min | 50 | 10.6 ± 0.2 | 179 ± 0.2 |
| | | 100 | 10.8 ± 0.2 | 186 ± 0.3 |
| | | 200 | 11.2 ± 0.1 | 190 ± 0.1 |

**Table 8.**
*Hardness and young modulus values of LSM-8YTZP material sintered by SPS.*

for the LSM material, they are observed to be 4.5 to be lower, i.e., the 8YTZP zirconia significantly inhibits the grain growth of the LSM.

The FESEM microstructures at different magnitudes of LSM-8YTZP composites sintered at 1200°C with a heating rate of 200°C/min, as shown in **Figure 7**. A homogeneous microstructure is observed. The small zirconia grains are located at the grain boundary inhibiting the growth of LSM grains.

The mechanical hardness and Young's modulus values can be seen in **Table 8**. Hardness values above 11 GPa and Young's modulus values around 190 GPa have been achieved using a heating ramp of 200°C/min. Moratal et al [20] achieved values around 6.0 GPa by sintering this LSM-8YTZP composite by conventional sintering at 1200°C–2 h, heating ramps of 10°C/min. In this study it was found that using the SPS technique almost twice the values of the conventional method were achieved. Another important factor to take into account is that by using SPS, whose heating rate is 200°C/min, the sintering cycle is significantly reduced. Therefore, the high added value of these composites is a very positive factor for their potential industrial applications.

### *3.2.2 LSM-8YTZP powders modified by diffuse coplanar surface barrier discharge (DCSBD) and sintered by SPS*

The LSM-8YTZP starting powders were modified by the DCSBD technique, using the parameters mentioned in Section 2.2, and then sintered by SPS at 1200°C–2 min, using the same parameters as described in Section 3.2.1.

**Table 9** shows the values of average grain size and relative density corresponding to the LSM-8YTZP powders modified by DCSBD and sintered by SPS at 1200°C–2 min using different heating rates. As in the case of LSM powders surface-modified by the DCSBD technique, also with LSM-8YTZP powders, the density values increase to more than 95% in the composite sintered at a heating rate of 200°C/min.

| Material DCSBD | Sintering parameters | Heating rate (°C/min) | Grain size (μm) | Relative density (%) |
|---|---|---|---|---|
| LSM-8YTZP | 1200°C—2 min | 50 | 0.92 ± 0.03 | 94.3 ± 0.5 |
| | | 100 | 0.94 ± 0.02 | 94.8 ± 0.5 |
| | | 200 | 0.89 ± 0.02 | 95.8 ± 0.5 |

**Table 9.**
*Grain size and relative density values of LSM-8YTZP powders modified by DCSBD and sintered by SPS.*

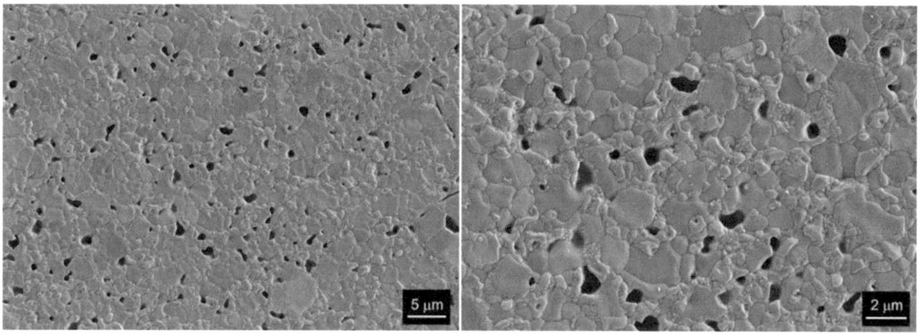


**Figure 8.**
*Microstructure of LSM-8YTZP composite with DCSBD treatment and SPS at 1200°C—2 min, heating ramp 200°C/min.*

The FESEM microstructures at different magnifications of the DCSBD-treated sample sintered at 1200°C with a heating rate of 200°C/min, as shown in **Figure 8**.

The grain size increases slightly in the composites with the modified powders compared to the non-modified ones, and average grain size values of around 0.90 μm are reached. A homogeneous microstructure can be observed, where besides the residual porosity, there is also grain removal due to the polishing process. Small particles of modified powder are situated at the grain boundaries, reinforcing the final material, as seen from the mechanical properties values in **Table 10**.

The LSM-8YTZP composite, sintered at 200°C/min, achieves hardness values of 12.5 GPa. This is very important when compared to the hardness value achieved by the DCSBD-modified LSM material sintered with the same parameters (8.3 GPa). On the one hand, it can be confirmed that the zirconia strengthens the LSM material, and the surface modification by the DCSBD technique also has a very positive effect. The dual effect of the composition and the surface modification increases by 60% both the hardness and the elastic modulus with respect to the untreated LSM material.

| Material DCSBD | Sintering parameters | Heating rate (°C/min) | Hardness (GPa) | Young modulus (GPa) |
|---|---|---|---|---|
| LSM-8YTZP | 1200°C—2 min | 50 | 11.0 ± 0.3 | 188 ± 0.3 |
| | | 100 | 11.8 ± 0.2 | 196 ± 0.1 |
| | | 200 | 12.5 ± 0.2 | 202 ± 0.3 |

**Table 10.**
*Hardness and young modulus values of LSM-8YTZP powders modified by DCSBD and sintered by SPS.*

### *3.2.3 LSM-8YTZP powders modified by atomic layer deposition (ALD) and sintered by SPS*

The LSM-8YTZP starting powders were modified by the ALD technique, using the parameters mentioned in Section 2.3, and then sintered by SPS at 1200°C—2 min, using the same parameters as in the previous section.

In **Table 11**, the values of average grain size and relative density corresponding to the LSM-8YTZP powders modified by ALD and sintered by SPS at 1200°C—2 min using different heating rates can be observed. These composites reach relative densities very close to those achieved by non-treated composites (**Table 7**), in the order of 93%. The grain size of the composites is similar to that achieved by the materials obtained by DCSBD treatment (**Table 9**).

The FESEM microstructures at different magnifications of the ALD-treated sample sintered at 1200°C with a heating rate of 200°C/min, as shown in **Figure 9**. It can be observed that in the FESEM micrograph obtained at low magnification, there are areas where agglomerates are present, the microstructure is not homogeneous. Residual porosity locates predominantly at the grain boundaries. Zirconia particles have a grain size of 0.3–0.4 μm. These particles inhibit the grain growth of the LSM, and therefore the ALD technique does not have a strong influence on this property. The higher magnification micrograph shows some grain removal due to the polishing phase.

The values of hardness and Young's modulus can be seen in **Table 12**. Hardness values of over 11.0 GPa and an average Young's modulus of around 195 GPa are achieved for all materials. These values are below those achieved when using the DCSBD technique but clearly above those of the LSM treated by ALD. The influence of the microstructure has a key role to play in the final properties.

| Material ALD | Sintering parameters | Heating rate (°C/min) | Grain size (μm) | Relative density (%) |
|---|---|---|---|---|
| LSM-8YTZP | 1200°C—2 min | 50 | 0.90 ± 0.04 | 92.1 ± 0.5 |
| | | 100 | 0.92 ± 0.07 | 92.5 ± 0.5 |
| | | 200 | 0.91 ± 0.05 | 93.4 ± 0.5 |

**Table 11.**
*Grain size and density values of LSM-8YTZP powders modified by ALD and sintered by SPS.*

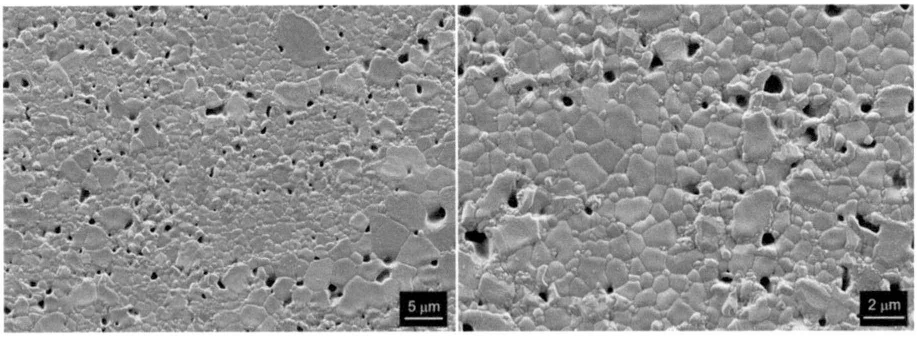


**Figure 9.**
*Microstructure of LSM-8YTZP composite with ALD treatment and SPS at 1200°C—2 min, heating ramp 200°C/min.*

| Material ALD | Sintering parameters | Heating rate (°C/min) | Hardness (GPa) | Young modulus (GPa) |
|---|---|---|---|---|
| LSM-8YTZP | 1200°C—2 min | 50 | 11.2 ± 0.2 | 190 ± 0.3 |
| | | 100 | 11.3 ± 0.2 | 192 ± 0.4 |
| | | 200 | 11.8 ± 0.3 | 201 ± 0.2 |

**Table 12.**
*Hardness and young modulus values of LSM-8YTZP powders modified by ALD and sintered by SPS.*

## 4. Conclusions

The addition of zirconia to the LSM material improved its final properties. The LSM material, being very porous, requires higher sintering temperatures to obtain a higher densification. By adding zirconia, the open porosity value of the material is reduced, thus allowing densification at lower working temperatures. In the study carried out, it was identified that with different sintering parameters, the composite material presents different behaviours that can favour some physical properties, depending on the application for which the material is used. In the case of the composite material, when used as a fuel cell, a better interconnectivity between the particles is necessary, so the presence of zirconia is essential. On the other hand, a level of porosity is necessary for the liberation of electrons in the reaction process.

With respect to the LSM and LSM-8YTZP materials, not treated and treated with the different DCSBD and ALD techniques, it can be concluded that the materials that offer the best results are those treated with the DCSBD technique. This technique causes a lower residual porosity in the material and contributes to inhibiting grain growth, which results in superior mechanical properties. The desired improvements are less significant when using the ALD technique.

However, it is important to take into account for future work that the sintering processes in this work were carried out only by SPS, so the modification of the surface may have different results in other non-conventional sintering methods such as microwave, which could be beneficial for the final properties of the materials.

## Acknowledgements

Grants PID2021-128548OB-C21 and PID2021-124521OB-100 funded by MCIN/AEI/10.13039/501100011033 and “ERDF A way of making Europe”. Grant CIGE/2021/076 funded by Generalitat Valenciana.

## Author details

Amparo Borrell[1]*, Rut Benavente[1], René M. Guillén[1], María D. Salvador[1], Vaclav Pouchly[2], Martina Ilcikova[3], Richard Krumpolec[3] and Rodrigo Moreno[4]

1 Instituto de Tecnología de Materiales (ITM), Universitat Politècnica de València, Valencia, Spain

2 Central European Institute of Technology (CEITEC), Brno, Czech Republic

3 Department of Physical Electronics, Masaryk University, Brno, Czech Republic

4 Instituto de Cerámica y Vidrio, CSIC, Madrid, Spain

*Address all correspondence to: aborrell@upv.es

## References

[1] Zhao S, Wang Y, Zhang Y, Bai J, Zhang Y, Wang S, et al. Enhancement of the redox reactions of the La0.8Sr0.2MnO3 catalyst by surface acid etching: A simple synthesis strategy to high-performance catalysts for methane combustion. Fuel. 2023;**345**: 128258

[2] Vinchhi P, Khandla M, Chaudhary K, Pati R. Recent advances on electrolyte materials for SOFC: A review. Inorganic Chemistry Communications. 2023;**152**: 110724

[3] Kima HW, Kim HJ, Yun KS, Peck DH, Moon J, Yu JH. Electrochemical characteristics of limiting current sensors with LSM-YSZ and LSM-CGO-YSZ composite electrodes. Ceramics International. 2023;**49**:21521-21529

[4] Li N, Sun L, Li Q, Xia T, Huo L, Zhao H. Novel and high-performance (La,Sr)MnO3 based composite cathodes for intermediate-temperature solid oxide fuel cells. Journal of the European Ceramic Society. 2023;**43**:5279-5287

[5] Sahini MG, Lupyana SD. Perspective and control of cation interdiffusion and interface reactions in solid oxide fuel cells (SOFCs). Materials Science & Engineering B. 2023;**292**:116415

[6] Borrell A, Salvador MD. Advanced ceramic materials sintered by microwave technology. Sintering Technology - Method and Application. Ed. Intech Open. 2018;**1**:3-24. DOI: 10.5772/intechopen.78831

[7] Manishanma S, Dutta A. Synthesis and characterization of nickel doped LSM as possible cathode materials for LT-SOFC application. Materials Chemistry and Physics. 2023;**297**:127438

[8] Budác D, Milos V, Carda M, Paidar M, Fuhrmann J, Bouzek K. Prediction of electrical conductivity of porous composites using a simplified Monte Carlo 3D equivalent electronic circuit network model: LSM-YSZ case study. Electrochimica Acta. 2023;**457**: 142512

[9] Gupta N, Mallik P, Lewis MH, Basu B. Improvement of toughness of Y-ZrO2: Role of dopant. Distribution. 2004;**268**: 817-820

[10] Moratal S, Guillén R, Salvador MD, Benavente R, Peñaranda-Foix FL, Moreno R, Borrell A. Comparative study of mechanical and microestructural properties of LSM ceramics obtained by E-field and H-field, AMPERE 2021 - 18th International Conference on Microwave and High Frequency Heating, Conference Paper, Code 176575

[11] Čech J, Hanusová J, Sťahel P, Černák M. Diffuse coplanar surface barrier discharge in artificial air: Statistical behaviour of microdischarges. Open Chemistry. 2015;**13**:528-540

[12] Skácelová D, Danilov V, Schäfer J, Quade A, Sťahel P, Černák M, et al. Room temperature plasma oxidation in DCSBD: A new method for preparation of silicon dioxide films at atmospheric pressure. Materials Science and Engineering B. 2013;**178**: 651-655

[13] Prysiazhnyi V, Vasina P, Panyala NR, Havel J, Cernak M. Air DCSBD plasma treatment of Al surface at atmospheric pressure. Surface and Coatings Technology. 2012;**206**: 3011-3016

[14] Pouchlý V et al. Improved microstructure of alumina ceramics

prepared from DBD plasma activated powders. Journal of the European Ceramic Society. 2019;**39**:1297-1303

[15] Černák M, Černáková L, Hudec I, Kováčik D, Zahoranová A. Diffuse Coplanar Surface Barrier Discharge and its applications for in-line processing of low-added-value materials. European Physical Journal: Applied Physics. 2009; **47**(2):1-6. DOI: 10.1051/epjap/2009131. hal-00497324

[16] Justin Kunene T, Kwanda Tartibu L, Ukoba K, Jen T-C. Review of atomic layer deposition process, application and modeling tools. Materials Today Proceedings. 2022;**62**:S95-S109

[17] ASTM C373–14. Standard Test Method for Water Absorption, Bulk Density, Apparent Porosity, and Apparent Specific Gravity of Fired Whiteware Products, ASTM C373–88. 88 (1999) 1–2

[18] ASTM E112. Standard Test Methods for Determining Average Grain Size E112–10, ASTM E112–10. 96 (2010) 1–27

[19] Oliver WC, Pharr GM. An improved technique for determining hardness and elastic modulus using load and displacement sensing indentation experiments. Journal of Materials Research. 1992;**19**:1564-1583

[20] Moratal S, Benavente R, Salvador MD, Peñaranda-Foix FL, Moreno R, Borrell A. Microwave sintering study of strontium-doped lanthanum manganite in a single-mode microwave with electric and magnetic field at 2.45 GHz. Journal of the European Ceramic Society. 2022;**42**: 5624-5630

Section 2

# Ceramic Materials for Multifunctional Applications

Chapter 4

# Permanent Magnets Based on Hard Ferrite Ceramics

*Cecilia Granados-Miralles, Matilde Saura-Múzquiz and Henrik L. Andersen*

## Abstract

Permanent magnets are integral components in many of the modern technologies that are critical for the transition to a sustainable society. However, most of the high-performance ($BH_{max}$ > 100 kJ/m$^3$) permanent magnets that are currently employed contain rare earth elements (REE), which have long been classified as critical materials with a high supply risk and concerns regarding pollution in their mining. Therefore, suitable REE-lean/free magnets must be developed in order to ensure the sustainability of clean energy generation and electric mobility. The REE-free hexagonal ferrites (or hexaferrites) are the most used permanent magnets across all applications, with an 85 wt.% pie of the permanent magnet market. They are the dominant lower-grade option ($BH_{max}$ < 25 kJ/m$^3$) due to their relatively good hard magnetic properties, high Curie temperature (>700 K), low cost and good chemical stability. In recent years, the hexaferrites have also emerged as candidates for substituting REE-based permanent magnets in applications requiring intermediate magnetic performance (25–100 kJ/m$^3$), due to considerable performance improvements achieved through chemical tuning, nanostructuring and compaction/sintering optimization. This chapter reviews the state-of-the-art sintering strategies being investigated with the aim of manufacturing hexaferrite magnets with optimized magnetic properties, identifying key challenges and highlighting the natural future steps to be followed.

**Keywords:** permanent magnets, hard ferrites, hexaferrites, ceramic magnets, rare earth-free magnets, $SrFe_{12}O_{19}$, $BarFe_{12}O_{19}$

## 1. Introduction

### 1.1 Classification of magnetic materials

The magnetism of magnetic materials arises at the atomic scale and is influenced by characteristics spanning several orders of magnitude (see **Figure 1a**). In the atoms of most compounds, the electrons exist in pairs with opposite spins that cancel out each other's magnetic moment. However, some elements or ions have unpaired electrons, whose spin and orbital motion cause them to exhibit a magnetic field, giving the atom a magnetic moment. The organization of these magnetic atoms in the atomic structure of the material determines its magnetic properties.

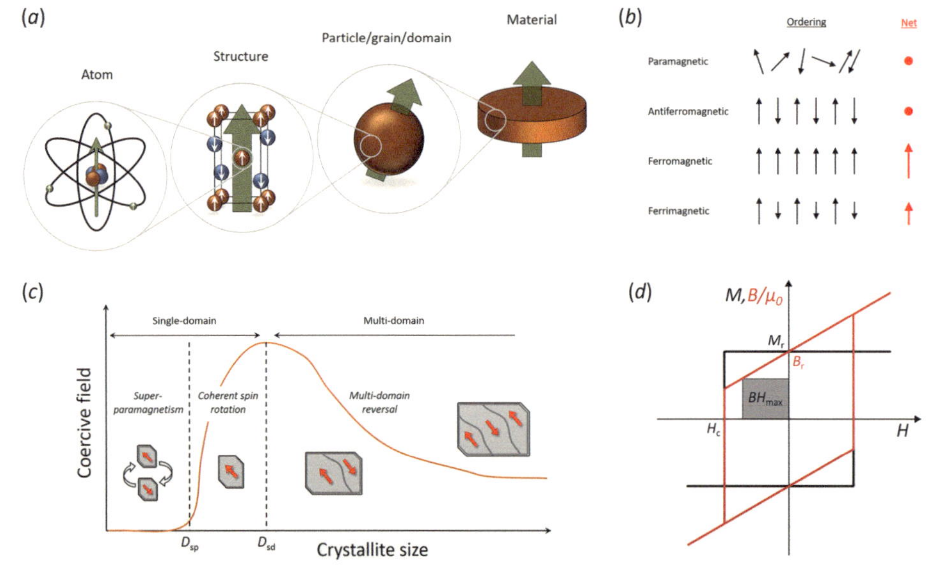


**Figure 1.**
*(a) Illustration of the multiscale origin of the magnetism in magnetic materials. (b) Schematic illustration of main magnetic ordering types and the resulting net zero field magnetization. (c) Size dependency of the coercivity. (d) Hysteresis curve of an ideal permanent magnet.*

**Figure 1b** shows a schematic illustration of the main types of magnetic ordering. In paramagnetic materials, the atomic magnetic moments are randomly oriented leading to no net magnetization and a relatively weak attraction to an external magnetic field. Antiferromagnetic materials are magnetically ordered, but also exhibit zero net magnetization due to an antiparallel organization of equal atomic magnetic moments. However, in ferro- or ferrimagnetic materials (below their Curie temperature, $T_c$, which is the critical temperature above which thermal fluctuations lead to the material being paramagnetic), the magnetic atoms are organized in a way that leads to a net magnetization along a certain direction (magnetic easy axis) in the structure, and it is these types of materials that are used for permanent magnets (PMs).

The ferro/ferrimagnetic materials are generally categorized as either "soft" or "hard" depending on their resistance to demagnetization. This is evaluated in terms of the coercive field (or coercivity, $H_c$), which is the external magnetic field required to reset the magnetization of the material. Magnetically soft materials are easily (de)magnetized by an external magnetic field (typically defined as $H_c < 10$ kA/m), and their magnetization is therefore often temporary, while hard (or permanent) magnetic materials have a high resistance to demagnetization ($H_c > 400$ kA/m) and once magnetized they can therefore sustain a magnetic field indefinitely [1]. The coercivity of a material is determined in part by the intrinsic magnetocrystalline anisotropy of the crystal structure as well as by microstructural (extrinsic) effects such as crystallite size or structural defects, which influence the formation (nucleation and growth) of magnetic domains in the material. For most magnetic materials, $H_c$ is found to increase as the crystallite size is reduced, reaching a maximum value at the critical single-domain size (see **Figure 1c**).

Another key property of a magnetic material is its remanence field ($B_r$ or $M_r$), which is the spontaneous magnetic flux density or magnetization exhibited by the material in zero external field conditions. **Figure 1d** shows a schematic illustration of the external magnetic field ($H$)-dependent flux density ($B$) and magnetization ($M$) curves, commonly called hysteresis curves, of an ideal permanent magnetic material. As illustrated, it is the combination of these two parameters, i.e., the coercivity (magnetic stability) and remanence (spontaneous magnetization), that ultimately determines the magnetic strength of the magnet. This magnetic strength is quantified by the so-called maximum energy product ($BH_{max}$), defined by the area of the largest possible rectangle that fits under the $BH$ curve in the second quadrant, which measures the potential energy stored in the stray field of the magnet [2].

**Figure 1d** shows the magnetic hysteresis of an ideal permanent magnet, in which all magnetic spins are perfectly aligned (and therefore, $M_r = M_s$), but in real magnets, the remanence value is smaller than the saturation (i.e., $M_r < M_s$). It follows that, as the $M_r$ value approaches $M_s$, the loop turns more squared, and in turn, $BH_{max}$ is maximized. Hence, the squareness and magnetic alignment are often measured in terms of in $M_r/M_s$ ratio [3], which is another of the key parameters to be improved for permanent magnets.

### 1.2 Materials for permanent magnets: Current status

Magnetic materials have the unique ability to directly interconvert between electrical and mechanical energy. A moving magnet can induce an electric current to generate electrical energy, and oppositely, an electric current can be used to generate a magnetic field and exert a magnetic force. These electromagnetic properties underpin the operation of electric generators and motors, making magnetic materials critical for the transition toward an environmentally friendly and sustainable future [2]. As a result, the worldwide permanent magnet market is expected to reach $39.71 billion by 2030, according to the 8.6% compound annual growth rate (CAGR) forecast in the last Grand View Research report [4].

**Figure 2a** illustrates the relative performance in terms of $BH_{max}$ and $H_c$ for the most important families of commercial PM materials, including AlNiCo alloys, hard ferrites ceramics, $Nd_2Fe_{14}B$ and $SmCo_5$. The high-performance ($BH_{max} > 100$ kJ/m$^3$) permanent magnet market is currently dominated by the rare earth element (REE)-containing materials $Nd_2Fe_{14}B$ (strongest magnet) and $SmCo_5$ (best high-temperature performance) due to their superior energy products [1, 6], which is a critical parameter for the performance in applications where miniaturization is a major driving force (e.g., electric vehicle motors, direct-drive generators, electro-acoustic devices, accessory electric motors, mobile phones, sensors, portable electronics, etc.). Unfortunately, the use of REE-based materials entails various problems. The compounds rely on scarce REE such as neodymium, samarium or dysprosium, which are classified as critical raw materials, not only owing to their supply risk and price volatility but also to the harmful environmental impact of their extraction [7]. China has been the undisputed leader in REE mining and production for the last 40 years [8], and despite other countries attempting to gain ground, today, China still accounts for more than 60% of the world REE production [9]. Consequently, over the last 20 years, geopolitical circumstances have often led to erratic price fluctuations. Furthermore, the cobalt used in $SmCo_5$ magnets is another problematic element. The supply chains for the bulk part (> 50%) of the cobalt used in advanced materials can be traced back to the cobalt mines in the Democratic Republic of the Congo, where artisanal miners (including thousands of children) work under extremely hazardous conditions [10]. As a consequence, the development of REE-poor or REE-free alternatives has long been an important research topic in the PM field.

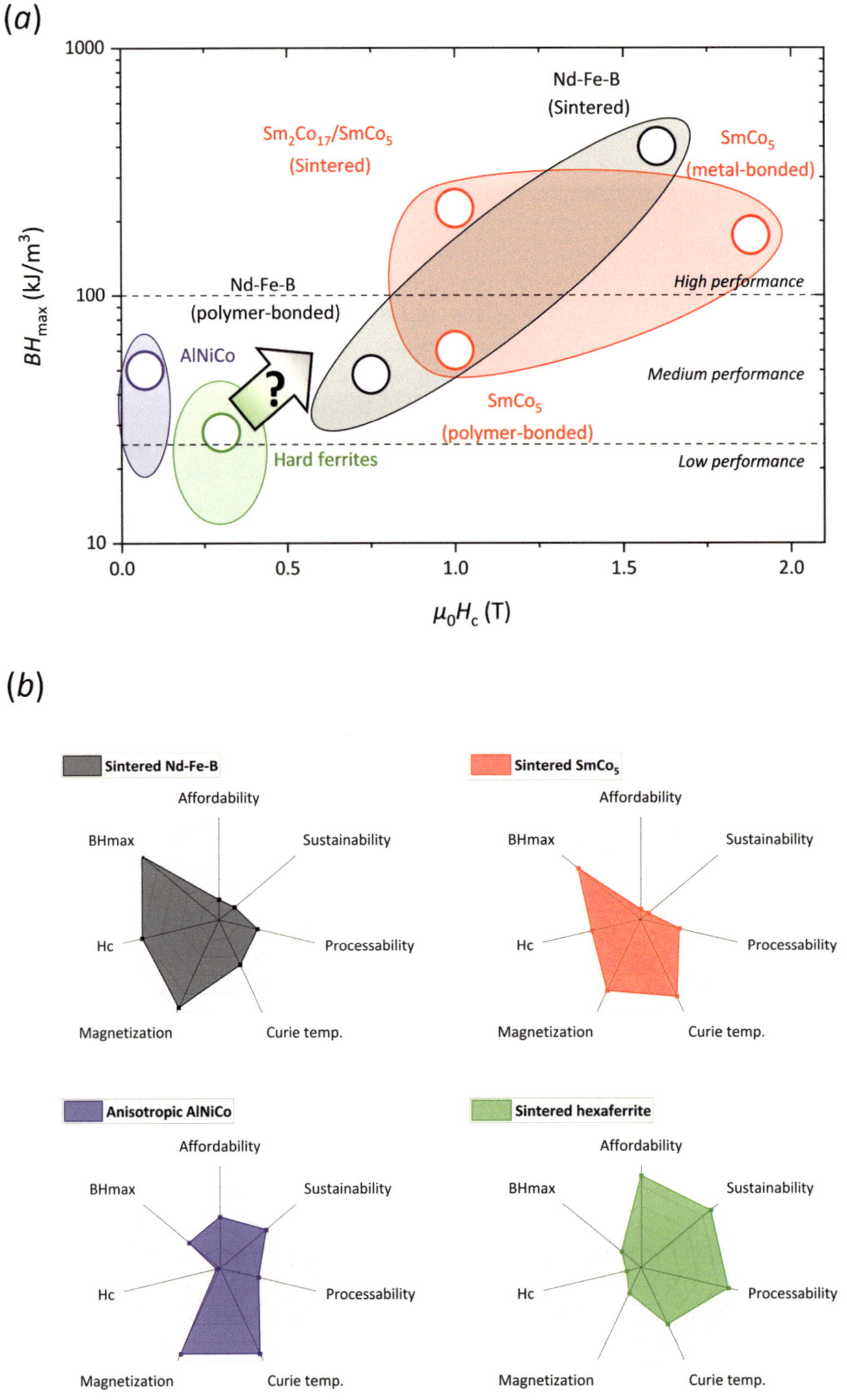


**Figure 2.**
*(a) Diagram of $BH_{max}$ vs. coercive field for the main families of commercially available hard magnetic materials. (b) Radar plots of key extrinsic properties of sintered $Nd_2Fe_{14}B$, sintered $SmCo_5$, anisotropic AlNiCo and sintered hexaferrite magnets. Figures based on values from [5].*

Although the undisputed strength of REE magnets is necessary for the highest-performance applications, there are many other applications that are less demanding in terms of magnetic strength, where a compromise (see **Figure 2b**) must be made between other factors such as price, stability, processability, etc. [11]. At this end of the spectrum, hard ferrite magnets have long been the material of choice for lower-grade applications (< 25 $kJ/m^3$). However, as illustrated by the arrow in **Figure 2a**, a considerable performance gap exists in the intermediate performance range between the cheaper AlNiCo and hard ferrite PMs and REE PMs. Consequently, for many applications, it is often necessary to use an expensive and excessively strong REE magnet, in lack of an intermediate alternative. Here, a modest performance improvement of lower-grade magnets would be sufficient to replace REE PMs while remaining within a weight range suitable for the application.

In this context, hexaferrites have long been considered good candidates for replacing REE magnets in the intermediate performance range due to their reasonably good performance, high Curie temperature (>700 K) and excellent chemical stability, which all come at a fraction of the cost of REE magnets [12, 13]. In fact, hard ferrites are the most produced magnetic material, despite their moderate performance compared to REE magnets [14]. In 2013 they were reported to account for 85% of the total PM market by manufactured mass, although they only represented 50% of the market by sales [15].

While recent studies have demonstrated new approaches to improve magnetic properties of hard hexaferrite powders (e.g., nanostructuring [16–19], chemical substitution [20–23], exchange spring composites [19, 24, 25]), manufacturing dense sintered pellets of sufficient structural integrity without degrading the optimized properties has proven a key challenge. In practice, this prevents the replacement of expensive and unsustainable REE PMs in a range of applications and is the reason why hard ferrites still generate great scientific interest [26]. The present chapter aims to summarize the most relevant recent achievements and progress in the field, as well as key challenges encountered during the fabrication and sintering of dense ferrite magnets.

## 2. Hard ferrites: M-type hexaferrites

### 2.1 Crystal and magnetic structure

The so-called hexaferrites, hexagonal ferrites or simply hard ferrites, are a family of ternary or quaternary iron oxides with hexagonal crystal lattice of long unit cell *c*-axis (≈23–84 Å) [26]. Of the materials in the hexaferrite family, the M-type hexaferrites have been widely used for application as permanent magnets. With the chemical formula $MFe_{12}O_{19}$ ($M$ = $Sr^{2+}$ or $Ba^{2+}$), the Sr. and Ba M-type ferrites (SrM and BaM) are isostructural and exhibit very similar magnetic characteristics. The compounds have a large uniaxial magnetocrystalline anisotropy and a magnetic easy axis along the crystallographic *c*-direction. This strong intrinsic anisotropy results in a high $H_c$, making them very resistant toward demagnetization (i.e., magnetically hard) and, therefore, attractive as PM materials.

**Figure 3** illustrates the crystal and magnetic structures of M-type hexaferrites. They display a hexagonal magnetoplumbite structure (space group $P6_3/mmc$) with a very anisotropic unit cell ($a \approx 5.9$ Å, $c \approx 23$ Å). $Fe^{3+}$ ions occupy interstitial positions in a hexagonal close-packed structure of $O^{2-}$ and $Sr^{2+}$ ($Ba^{2+}$) ions [26–28]. With two formula units per unit cell (64 atoms), SrM has a crystallographic density of

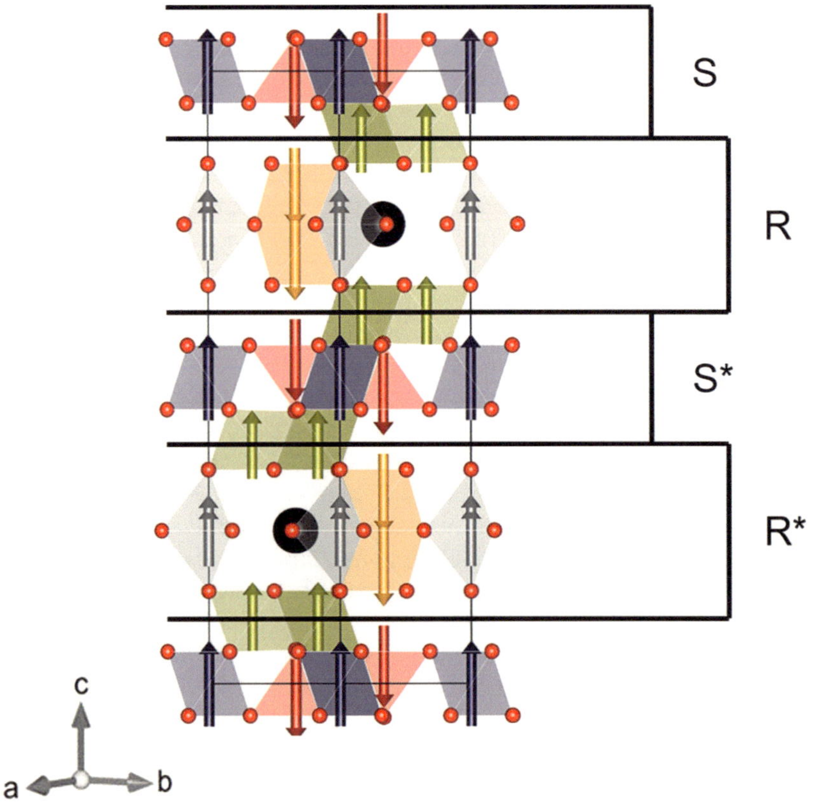


**Figure 3.**
*Crystal and magnetic structure of Sr. (Ba) hexaferrite. Black and red spheres represent $Sr^{2+}$ ($Ba^{2+}$) and $O^{2-}$ ions. Colored polyhedra illustrate the five different crystallographic sites of $Fe^{3+}$ and arrows symbolize the $Fe^{3+}$ magnetic spins.*

5.3 g/cm$^3$ (5.1 g/cm$^3$ for BaM) [29, 30]. The crystal structure may also be described in terms of stacking of simpler structural blocks (cubic S and hexagonal R blocks), which are in turn stacked onto similar blocks rotated 180° about the *c*-axis (S* and R* blocks, respectively) [28].

## 2.2 Magnetic properties

**Table 1** compares the intrinsic magnetic properties of SrM and BaM with that of other important magnetic compounds. The theoretical magnetic moments (at 0 K) of the hexaferrite crystal structures can be calculated from the ferrimagnetic ordering of the magnetic $Fe^{3+}$ ions in the structure (see arrows in **Figure 3**), yielding values of 20.6 $\mu_B$/molecule for SrM and 20 $\mu_B$/molecule for BaM [26, 32]. This results in fairly good saturation magnetization, $M_s$, and magnetic induction, $B_s$, values. The Curie temperature, $T_C$, of the M-type hexaferrites is more than 100°C above that of the much-used REE-based $Nd_2Fe_{14}B$ hard phase.

The large uniaxial anisotropy of the hexagonal lattice of SrM and BaM ($c/a$ = 3.9) causes a large magnetocrystalline anisotropy along the *c*-axis, which yields relatively high anisotropy constants, $K_1$ (see **Table 1**) [31, 33, 34] and a large theoretical maximum $H_c$ of 594 kA/m [26]. For a hypothetical fully dense and perfectly-oriented hexaferrite magnet, a theoretical maximum $BH_{max}$ of 45 kJ/m$^3$ has been estimated [1].

| | $M_s$ (Am²/kg) | $B_s$ (T) | $T_C$ (K) | $K_1$ (MJ/m³) |
|---|---|---|---|---|
| $Fe_{0.65}Co_{0.35}$ | 240 | 2.45 | 1210 | 0.018 |
| Fe | 217 | 2.15 | 1044 | 0.048 |
| AlNiCo5 [1] | 159 | 1.40 | 1210 | 0.68* |
| $CoFe_2O_4$ [5] | 75 | 0.5 | 793 | 0.27 |
| $BaFe_{12}O_{19}$ | 72 | 0.48 | 740 | 0.33 |
| $SrFe_{12}O_{19}$ | 72 | 0.48 | 746 | 0.35 |
| $Nd_2Fe_{14}B$ | 165 | 1.61 | 588 | 4.9 |
| $SmCo_5$ | 100 | 1.07 | 1020 | 17.2 |
| $Sm_2Co_{17}$ | 118 | 1.25 | 1190 | 4.2 |

**Shape anisotropy.*

**Table 1.**
*Intrinsic magnetic parameters at room temperature (RT) for some representative soft and hard magnetic phases. Data extracted from [31] unless otherwise stated.*

## 3. Sintered hard ferrite permanent magnets

Toward the effective implementation of permanent magnets into a device, the material in powder form has to be compacted into dense, mechanically stable and magnetically oriented pieces (i.e., magnets). This conforming/densification process (called sintering) generally involves applying elevated pressures and/or temperatures to the material in powder shape [35]. As for most other materials, the mechanical properties of the sintered piece rely on a high density. However, the importance of achieving a highly dense magnet is enhanced for PMs, since the magnetic performance ($BH_{max}$) is measured per volume unit, and hence, it is directly proportional to the density.

The high sintering temperatures often end up undesirably altering the functional properties of the starting material, and therefore, great efforts are dedicated to both (i) adapting the sintering methods to the specific material of interest and (ii) developing novel sintering strategies that lower the working temperatures, aiming at minimizing the damage [35]. In the particular case of hexaferrites, a common problem is the formation of hematite ($\alpha$-$Fe_2O_3$) as a side phase. This iron oxide is very prone to appear, as a result of its high stability, and causes a decrease of saturation magnetization due to the antiferromagnetic nature of the phase. Fortunately, it has been shown that $\alpha$-$Fe_2O_3$ can be avoided when the starting $MFe_{12}O_{19}$ powders have the right $M$:Fe stoichiometry, yielding $M_s$ values approaching the expected ≈70 Am²/kg [26]. In contrast, limiting grain growth to circumvent the detrimental impact on $H_c$ has proven more challenging [36]. Thus, M-type ferrites in powder form often present coercivities far below the theoretical value, and the situation worsens for sintered pieces (see Table 7 in ref. [26] for an extensive sample record). Owing to this, sintered hexaferrite magnets are generally inferior to the theoretically achievable 45 kJ/m³ [1], although specific studies have managed to come fairly close to this value.

Another important aspect in the sintering of PMs is the magnetic alignment of the constituent particles and domains in the material. The magnetic particles may (or may not) be magnetically aligned, resulting in anisotropic (isotropic if not aligned) magnets. The greater the magnetic alignment, the more the $M_r$ value approaches $M_s$, yielding a more square-shaped $MH$ curve (as illustrated by the black curve in

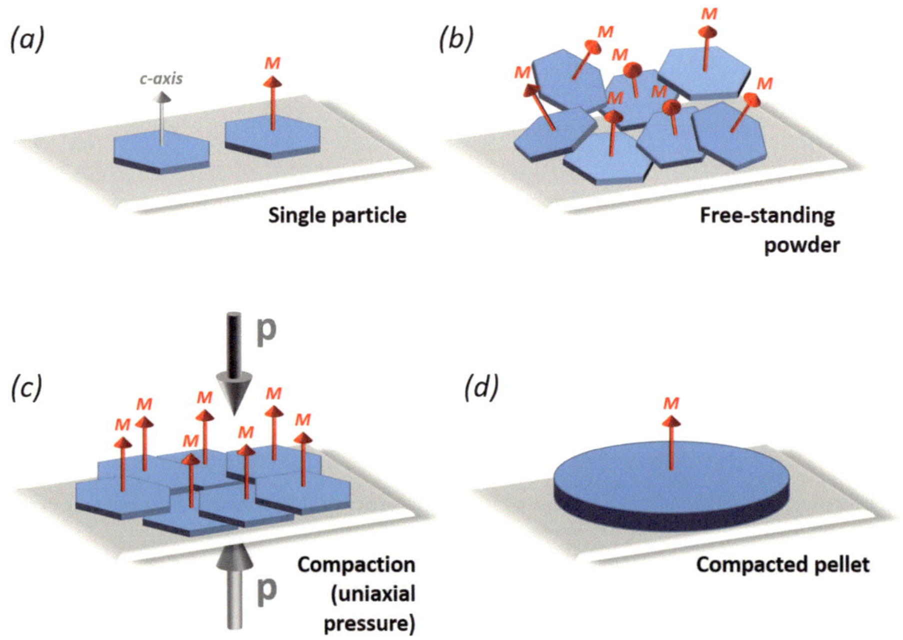


**Figure 4.**
*$MFe_{12}O_{19}$ particles, displaying typical hexagonal platelet shape with the easy axis of magnetization normal to the platelet plane (and parallel to the crystallographic c-axis). This shape favors magnetic (and crystallographic) alignment upon application of uniaxial pressure. Adapted with permission from [13].*

**Figure 1c**), thereby maximizing $BH_{max}$. Thus, the $BH_{max}$ of mass-produced isotropic M-ferrite magnets is around 10 kJ/m$^3$, while the anisotropic kind ranges from 33 to 42 kJ/m$^3$ [37–40].[1] The magnetic alignment has been traditionally carried out by application of an external magnetic field [26, 41], although recently patented methods have succeeded in suppressing the external field by taking advantage of the shape of the particles [42, 43]. Notably, the M-type ferrites are prone to form platelet-shaped particles, with magnetization direction parallel to the platelet normal vector (see **Figure 4a**). As illustrated in the figure, the platelet shape of the particles favors magnetic (and crystallographic) alignment upon compaction.

During the last decades, different sintering strategies have been investigated aiming at maximizing both the magnetic alignment (boosting $B_r$ and $M_r/M_s$) and the $H_c$ on the sintered material. Lately, efforts have also been devoted to making the processes greener and increasing recycling rates. The following sections intend to offer an overview of the pros and cons of each of the alternatives.

### 3.1 Conventional sintering

Hexaferrites were first developed as a PM material by researchers at the Philips Research Laboratories in 1950s. In 1952, Went et al. prepared a Ba-ferrite magnet with a good $H_c$ value (≈240 kA/m), although a limited $B_r$ derived from its isotropic nature (0.21 T) yielded a modest $BH_{max}$ of 6.8 kJ/m$^3$ [44, 45]. Two years later, Stuijts et al. developed a conventional sintering (CS) strategy to produce anisotropic BaM magnets

[1] Ferrite magnets with higher *BH*max values are available commercially (up to 44 kJ/m), but in those cases the material is doped with, e.g., La or Co. [39, 40].

with $BH_{max}$ up to 28 kJ/m$^3$ [41], which is essentially the method used nowadays to make sintered ferrite magnets industrially. In brief, a sludge of BaM powders and water is compacted while being held in an external magnetic field, producing a consolidated piece (still poor in density) which is subsequently sintered at temperatures above 1100°C to promote densification. Stuijts et al. explored sintering temperatures between 1250 and 1340°C and noted that increasing the temperature maximizes the density and the magnetic alignment (and therefore $B_r$) but at the cost of decreasing $H_c$, as a consequence of the grain growth promoted by the elevated temperatures. This problem, encountered already in 1954, has been the subject of extensive research since.

As mentioned earlier, structural characteristics such as crystallite size, size distribution, and crystallite morphology can largely affect the coercivity of ferrite magnets. In particular, the highest $H_c$ values are attained for crystallite sizes close to the critical single-domain size defined earlier [34, 45, 46]. The difficulty not only lies in being able to produce particles of a specific size in a controlled manner, but it begins with determining what this critical size is for a specific material. For isotropic SrM crystallites, the critical single-domain size has been estimated to be around 620–740 nm [16, 47]. However, the experimentally reported crystallite/particle single-domain sizes of SrM span from 30 nm all the way up to 830 nm [47]. This is due to the high influence of particle morphology in the attained coercivity, as well as to the different characterization methods used to determine the reported size (i.e., particle vs. crystallites, number vs. volume weighted, etc.). A study by Gjørup et al. showed that a much smaller critical single-domain size is obtained for highly anisotropic crystallites, and therefore not only the overall size but also the aspect ratio of anisotropic SrM crystallites should be considered when trying to maximize $H_c$ [47].

Notably, reducing the size of the starting powders does not necessarily yield better coercivities, as the grain growth upon sintering seems to be even greater when dealing with materials of smaller particle sizes [48–50]. El Shater et al. sintered nanometric BaM (100–200 nm) at 1000 and 1300°C, producing average particle sizes of 0.537 and 16.35 μm, respectively, with the consequent drop in coercivity (from 271 to 56 kA/m) and the $M_r$ gain [51]. Therefore, the choice of sintering temperature must be a compromise between minimizing grain growth (to maximize $H_c$) and maximizing densification (and, in turn, $M_r$).

A common approach for limiting grain growth has been the use of sintering additives. Kools proposed a mechanism through which $SiO_2$ would prevent the growth of SrM grains during sintering and proved the effect for a range of $SiO_2$ concentrations (0.36–1.44 wt.%) [52, 53]. Beseničar et al. reported that, besides limiting the growth, $SiO_2$ induces some ordering of the SrM particles, resulting in very anisotropic magnets with high relative density (97%) and satisfactory magnetic properties ($B_r \approx 0.39$ T, $H_c \approx 340$ kA/m) [54]. Kobayashi et al. determined the optimal $SiO_2$ concentration to be between 1 and 1.8 wt.%, showing a detrimental effect on $H_c$ for greater $SiO_2$ additions [55]. Guzmán-Mínguez et al. reported the appearance of ≈20 wt.% $\alpha$-$Fe_2O_3$ as a secondary phase for $SiO_2$ concentrations>1 wt.% [56].

CaO has been reported to favor densification, and therefore, it has also been explored as a sintering additive for hexaferrites, in this case, with the aim of boosting $M_r$, although at the expense of aggravating the grain growth effect [46, 55, 57]. In this context, the combined use of both additives has also been investigated. Lee et al. reported a decent $BH_{max}$ of 29.4 kJ/m$^3$ when adding 0.6 wt.% $SiO_2$ and 0.7 wt.% CaO, but neither remanence nor coercivity were terrific ($B_r$ = 0.36 T, $H_c$ = 281 kA/m) [58]. Töpfer et al. fabricated a very dense SrM magnet (98%) with a notable $B_r$ value of 0.42 T by incorporating 0.25 wt.% of $SiO_2$ and 0.25 wt.% CaO, although a moderate coercivity value of 282 kA/m only allowed for a $BH_{max}$ = 32.6 kJ/m$^3$ [59]. Huang et al. tested the combined addition of $CaCO_3$, $SiO_2$ and $Co_3O_4$ (1.1, 0.4 and 0.3 wt.%, respectively), managing a remarkable $BH_{max}$ of 38.7 kJ/m$^3$, owing to an exceptional remanence (0.44 T) and despite a modest coercivity (264 kA/m) [60].

Slightly superior magnetic parameters ($B_r$ = 0.44 T, $H_c$ = 328 kA/m, $BH_{max}$ = 37.6 kJ/m$^3$) have been obtained from a two-step sintering (TSS) method adapted to SrM by Du et al. [61]. Here, the powders were cold-pressed as usual CS, but the subsequent thermal cycle used for sintering was slightly more elaborate: after the first high-temperature step, in which the maximum temperature (1200°C) is maintained for only 10 min, a longer (2 h) heating step at 1000°C provides for full densification of the SrM magnet [61]. The scanning electron micrograph (SEM) in Figure 6e from ref. [61] illustrates the confined grain size, the high density and high degree of alignment, justifying the good magnetic performance. A more recent work by Guzmán-Mínguez et al. [62] combined a TSS approach with the addition of 0.2% PVA and 0.6% $SiO_2$, realizing great control of the grain growth at 1250°C (see **Figure 5**), although the obtained magnetic properties were not as good as the ones previously reported by Du et al.

### 3.2 Spark plasma sintering

In the 1990s, a new commercial apparatus based on resistive sintering, called spark plasma sintering (SPS) was developed by Sumitomo Heavy Industries Ltd. (Japan) [63]. The SPS method is based on the use of an electrical current and a uniaxial mechanical pressure under low atmospheric pressure, to simultaneously heat and compact a powder sample [64]. The starting powders are typically loaded in a graphite die, which is placed between two electrodes in a water-cooled vacuum chamber. A uniaxial pressure is applied to the die while passing a DC electrical current through, which heats up the sample due to the Joule effect (see Figure 1 in ref. [65] for a typical SPS setup). The inventors of the system claimed the generation of plasma to take place, thus leading to the technique's name. However, although it is generally accepted that plasma may be generated between particles due to electrical discharges, there is no conclusive experimental evidence of such an occurrence [64]. Therefore, SPS is sometimes referred to by alternative names, such as the field-assisted sintering technique (FAST). The simultaneous application of temperature and pressure can also be obtained by conventional hot pressing (HP). However, in SPS and HP, heat is produced and transmitted to the material in different ways. In conventional heating

**Figure 5.**
*SEM images of SrM pellet sintered at 1250°C by (a) Conventional sintering and (b) Two-step sintering. Reprinted from [62], copyright 2021, with permission from Elsevier.*

the powders are sintered by heating the entire container using external heating elements in a furnace. This leads to slow heating rates, long sintering times and waste of energy in heating up all the components. The SPS method, however, has allowed for increasing the heating rates, lowering the working temperatures and reducing the dwell times [66, 67]. These benefits make SPS a good alternative when the goal is to limit the grain growth during sintering [67], and potentially improve the obtained $H_c$ (and $BH_{max}$) values of sintered hexaferrite magnets.

Numerous investigations focusing on sintering hexagonal ferrites by SPS have been published in the last two decades. Obara et al. prepared fully dense SrM magnets by SPS at 1100°C and 50 MPa for only 5 min [65]. A fairly competitive $H_c$ of 325 kA/m was obtained by doping with $La_2O_3$ (1 wt.%) and $Co_3O_4$ (0.1 wt.%). Although the measured hysteresis loops were rather squared, the remanence value (0.32 T) was not sufficient to guarantee a noteworthy energy product ($BH_{max}$ = 18.3 kJ/m$^3$). Mazaleyrat et al. sintered BaM nanopowders with sizes below 100 nm and managed to hold grain growth and produce a $H_c$ of 390 kA/m [68], which even surpasses the value reported for the La and Co-doped material described above. Unfortunately, a deficient density (88%) degraded the $BH_{max}$ down to 8.8 kJ/m$^3$. Ovtar et al. sintered the same batch of 90 nm BaM nanoparticles by both CS and SPS, producing much smaller sizes through the second method [69]. Additionally, they realized that secondary phases ($Fe_3O_4$ and $\alpha$-$Fe_2O_3$) tend to form on the surface of the BaM SPS pellets, and tested different materials for the protective discs separating the sample from the graphite die (BN, Au and $\alpha$-$Al_2O_3$) concluding that $\alpha$-$Al_2O_3$ was the one performing best. The resulting density was rather low 82%, but the coercivity was adequate (350 kA/m). Stingatiu et al. attempted downsizing a μm-sized SrM material by a ball-milling step prior to consolidation through SPS [70]. The resulting density was satisfactory (90%), but unfortunately, ball-milling was seen to amorphized the surface of the SrM, which triggered the formation of secondary phases during SPS, this having a detrimental effect on the magnetic properties ($BH_{max}$ < 10 kJ/m$^3$).

Saura-Múzquiz et al. prepared nm-sized SrM powders by hydrothermal synthesis (HT) with hexagonal plate-like particles (such as those in **Figure 4**) with very small sizes; in some cases, the platelets were as thin as a single unit cell (i.e., <3 nm) [17]. These HT-synthesized SrM powders were consolidated by SPS yielding appropriate $H_c$ values of 301 kA/m. More importantly, the highly anisotropic shape of the particles provided for a pronounced magnetic alignment of the sintered SrM magnets, inherently occurring as a result of the simultaneous application of elevated temperature and uniaxial pressure, just as illustrated in **Figure 4**. Here, an $M_r/M_s$ ratio of 0.89 was reached without applying an external magnetic field neither before nor during sintering, yielding a $BH_{max}$ value of 26 kJ/m$^3$. **Figure 6a** shows the magnetic hysteresis of the HT powders and the corresponding SPS pellet, evidencing the squareness of the latter. Achieving magnetic alignment without a magnetic field is very convenient from an industrial point of view because it allows a full step to be removed from the manufacturing process (i.e., the magnetization), which simplifies the procedure, reduces costs and increases energy efficiency [42]. **Figure 6b** displays the powder X-ray diffraction (PXRD) data measured on both SrM powders and SPS pellets. Despite the very dissimilar appearance, Rietveld analysis demonstrates that both PXRD patterns are consistent with pure-phase $SrFe_{12}O_{19}$ although with notable differences in crystallite size and orientation. The highly anisotropic shape of the powders is visible from the sharpness of the *hkl*-reflections describing the crystallite on the platelet plane, such as (110) or (220), compared to the large broadening of those associated with the platelet thickness, e.g., (008), all this in agreement with much smaller sizes along the *c*-axis than on the *ab*-plane (i.e., thin platelets). Regardless of the difference in peak broadening, Bragg reflections of all orientations are present in the PXRD pattern measured for the SrM powders, demonstrating a random orientation of the crystallites. However, the very intense *hh*0 reflections are absent for the PXRD pattern recorded for the SPS pellet, while 00 *l* reflections (as well as others with high contribution from the *c*-crystallographic direction) are systematically intensified, thus indicating a marked preferred orientation of the platelets. As explained before, for M-type platelet-shaped particles, crystallite/particle alignment goes together with magnetic alignment. The crystallographic alignment was further studied based on pole figure measurements (**Figure 6c**), a slightly more complex diffraction measurement enabling quantification of the degree of orientation (**Figure 6d**).

Optimization of both the HT synthesis route [71] and the SPS protocol [18, 72], as well as the correlation of structural and magnetic properties, allowed reaching $M_r/M_s$ ratios as high as 0.95, although at the cost of reducing $H_c$ down to 133 kA/m, with which the $BH_{max}$ improvement was only moderate (29 kJ/m$^3$) [73]. However, performing a thermal treatment at 850°C after SPS was enough to reach a $BH_{max}$ = 36 kJ/m$^3$, value on the order of the highest-grade commercially available ferrite magnets [37–40], while avoiding the use of an external magnetic field. Applying this SPS protocol to SrM powders produced by synthesis methods other than HT did not yield such outstanding magnetic properties due to an inferior particle orientation degree and, hence, a poorer magnetic alignment [72, 74]. A newer study by Saura-Múzquiz et al. confirmed that the degree of magnetic alignment using this preparation method could be tuned by modifying the aspect ratio of the initial powders, reaching almost fully-aligned pellets ($M_r/M_s$ = 0.9) with densities above 90% [74]. Higher alignment leads to higher squareness and thus greater $M_r/M_s$ ratio and $BH_{max}$, but it is accompanied by a reduction in $H_c$ due to the inversely proportional relationship that exists between magnetization and coercive field. Nonetheless, by reducing the degree of alignment, they were able to obtain SrM magnets with a large $H_c$ of 412 kA/m, proving the potential of SPS to overcome the reduction of $H_c$ due to excessive crystallite growth.

Recently, Vijayan et al. reported the use of SPS not only for densification of ferrite powders but for the direct synthesis of aligned SrM magnets [75–77]. In this study,

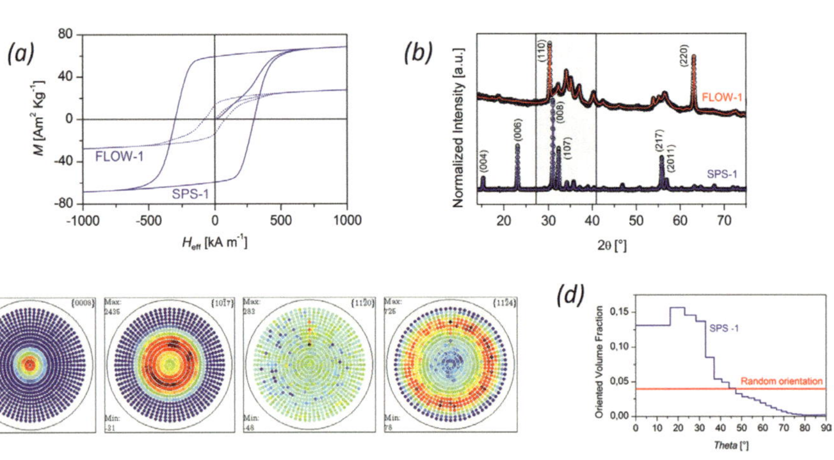


**Figure 6.**
*(a) Magnetic hysteresis loop of HT-synthesized SrM nanoparticles and corresponding SPS pellet. (b) PXRD data along with the Rietveld model of the same samples. (c) X-ray pole figure measurements and (d) Oriented volume fraction of SPS pellet. Reproduced from ref. [17] with permission from the Royal Society of Chemistry.*

SrM is synthesized directly during the SPS process using a precursor powder of antiferromagnetic six-line ferrihydrite (FeOOH) platelets mixed with $SrCO_3$. A low SPS temperature of ≈750°C was sufficient to drive the reaction between the six-line phase and $SrCO_3$ to produce $SrFe_{12}O_{19}$, while the anisotropic shape of the hydrothermally synthesized six-line phase ensured the alignment of the resulting SrM particles. Following this synthesis method, they were able to produce a dense SrM magnet with a $BH_{max}$ of 33(4) kJ/m$^3$, a $M_r/M_s$ of 0.93 and a $H_c$ of 247 kA/m.

### 3.3 Microwave sintering

In the field of hexaferrite research, microwaves (MWs) have mainly been used for synthesis purposes, although a few sintering attempts using MWs have also been reported [78–80]. In all of them, powders are initially cold-pressed, followed by a MW treatment, using frequencies in the GHz range, to sinter the piece [35]. In 1999, Binner et al. used MWs to sinter ferrite nanoparticles reporting a limited grain growth for nonagglomerated starting powders, although they failed to avoid cracks in the final sintered pieces [78]. Ten years later, Yang et al. succeeded in making 97% dense BaM magnets by MWs sintering [79]. They also managed to prevent the appearance of $\alpha$-$Fe_2O_3$ in the final material, although they did not succeed in preventing grain growth, in turn producing a rather low $H_c$ (<50 kA/m).

Recently, Kanagesan et al. tested fast heating rates (50°C/min) and short dwell times (10 min) to MW sinter some Sr-ferrite powders synthesized by sol-gel [80]. The MW sintering at 1150°C yielded a 95% SrM ceramic magnet with a fairly high $H_c$ of 445 kA/m. However, the $M_s$ value (50 Am$^2$/kg) was not outstanding, although the sample seems relatively pure from powder diffraction data. The $M_r$ value is also rather low (≈30 Am$^2$/kg), which is expected from the poor alignment of the $SrFe_{12}O_{19}$ particles observed in the corresponding SEM micrograph (see Figure 2 in ref. [80]).

### 3.4 Cold sintering process

In 2016, Guo et al. reported a new sintering strategy named cold sintering process (CSP), with which they were able to attain high densification degrees for a wide range of inorganic materials at temperatures ≤200°C, fabricating materials with functional properties comparable to those made by conventional high-temperature approaches [81]. For CSP, the ceramic powders are mixed with a small amount of aqueous solution, which partially dissolves the particle edges and facilitates diffusion and mass transport, aiding the sintering process, which in turn occurs at lower temperatures. Sintering at low temperatures is very attractive in general, as it reduces energy demands, making the process greener and more cost-efficient. This is especially interesting for M-ferrite magnets, as lower working temperatures are expected to minimize grain growth. The exact role of the solvent during CSP is still under discussion, but it is believed to induce the formation of an amorphous phase at the grain boundaries, which eases sintering and may also restrict grain growth [81].

To the best of our knowledge, there is only one research group that has tested CSP on hard ferrites. In 2021, Serrano et al. patented a CSP method that allows the fabrication of dense SrM magnets with magnetic properties in the order of medium-grade commercial ferrite magnets [43]. In the CSP method developed by Serrano, SrM powders are mixed with glacial acetic acid, and the wet mixture is subjected to a uniaxial pressure (≈400 MPa) while heated at 190°C [82]. After CSP, relative densities of about 85% are obtained, which can be driven up to 92% by subsequently treating the sintered piece at 1100°C for 2 h. This last sintering step also has a beneficial effect on the magnetic properties (see **Figure 7a**). In particular, $M_s$ at 5 T increases from 49.2 to 73.7 $Am^2/kg$, and $H_c$ goes from 119 to 223 kA/m. For the final product, a $M_r/M_s$ ratio of 0.68 was obtained. The density obtained by conventional sintering at 1100°C for 4 h (no solvent, no hot compression) was only 77%, and the magnetic properties were slightly inferior (see **Figure 7b**). Conventional sintering at 1300°C yielded higher density (97%) but very poor magnetic properties ($H_c$ = 48 kA/m and $M_r/M_s$ = 0.33) due to the dramatic grain growth caused by the high temperature (see bottom FE-SEM micrograph from **Figure 7b**).

Further investigations have been carried out using different organic solvents (i.e., oleic acid and oleylamine) and widening the pressure and temperature ranges explored (up to 270°C and 670 MPa) [83]. In all cases, the average grain size of the CSP ceramic magnet was about 1 μm (similar to the starting SrM powders), while similar conventional sintering processes typically yield average grain sizes above 3 μm [62].

With the aim of further improving the density and magnetic properties of CSP magnets, the addition of a small amount (10 wt.%) of nanometric SrM to the original micrometric SrM powders was tested, moderately increasing $H_c$ (239 kA/m) and $M_r/M_s$ (0.73), although the density value continued at 92% [84]. These numbers are

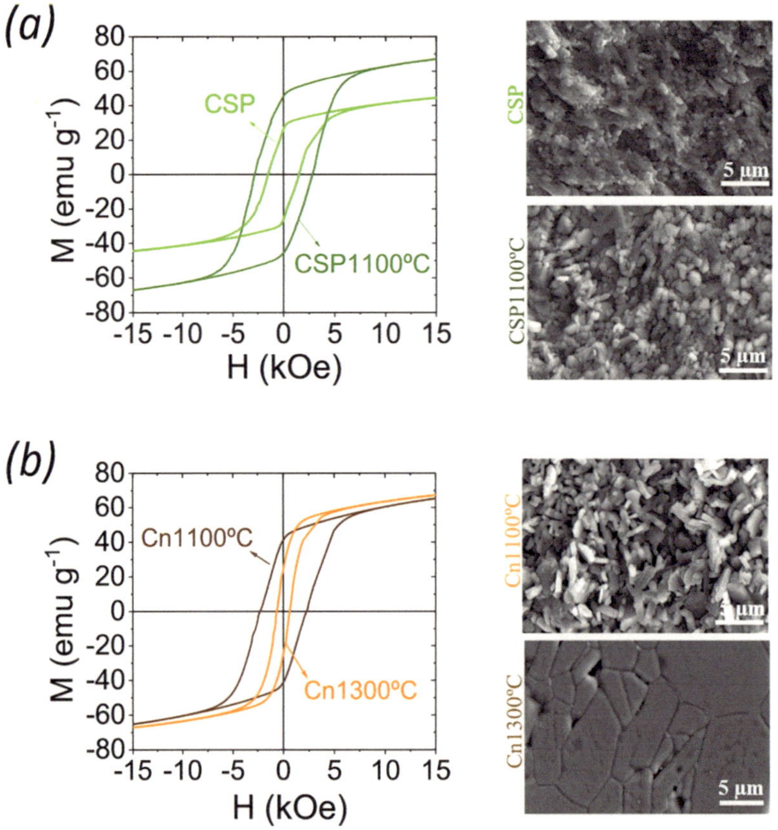


**Figure 7.**
*Magnetic hysteresis and FE-SEM correspond to $SrFe_{12}O_{19}$ magnets fabricated by (a) CSP at 190°C, CSP followed by annealing at 1100°C, (b) conventional sintering at 1100 and 1300°C. Reprinted from [82] with permission from Elsevier.*

competitive in the context of commercial SrM magnets. As an example, Hitachi's NMF-7C series display values of $H_c$ = 220–260 kA/m and $M_s$ = 68 $Am^2/kg$) [40].

## 4. Summary and perspective

In the present chapter, the main sintering approaches applied to manufacturing hard ferrite ceramic magnets have been reviewed. **Table 2** summarizes the properties of top $SrFe_{12}O_{19}$ magnets fabricated by the various discussed sintering strategies. Conventional sintering (CS) continues to be the quintessential industrial method for M-type hexaferrite PM fabrication, owing to its technical simplicity and relatively good resulting properties. However, this approach is highly inefficient, as most of the energy employed is irreversibly dissipated as heat [85]. Therefore, the search for more energy-efficient methods continues to be an active field of research.

Multiple studies have demonstrated that spark plasma sintering (SPS) allows the production of PMs with much higher $M_r/M_s$ ratios than CS. However, the increase in texture comes at the cost of reduction in $H_c$ values, which, therefore, still needs to be improved. As a result, magnets made using SPS end up displaying a similar performance ($BH_{max}$) to the best CS examples. Additionally, technical challenges hinder the replacement of CS by SPS in the industrial production of magnetic ferrites, since current SPS machines only allow producing relatively small pieces with very few specific shapes (typically cylindrical pellets).

Only a few attempts have so far been made to densify SrM by the relatively new cold sintering process (CSP) and therefore, there is still much to explore and optimize. However, the CSP has already allowed the preparation of hexaferrite magnets with magnetic properties comparable to medium-high grade commercial ferrites, while lowering the sintering temperature. This reduces energy consumption by about 9 kWh/kg, which leads to energy savings of ≈29% compared to the sintering methods employed industrially at present.

The results obtained by microwave sintering (MWs) have been very satisfactory in terms of both density and $H_c$, but the resulting $M_s$ and $M_r/M_s$ values are still insufficient to be commercially competitive. As with CSP, reports are scarce and further exploration is required.

Sintering has undergone significant innovation over the last decade [35], with the introduction of a number of new sintering technologies, such as flash sintering [86, 87], and various modified SPS methodologies, like flash SPS (FSPS) [88], deformable punch SPS (DP-SPS) [89], or cool-SPS [90]. As a result, there are more

| | $M_s$ ($Am^2/kg$) | $M_r/M_s$ | $M_r$ ($Am^2/kg$) | $H_c$ (kA/m) | $BH_{max}$ ($kJ/m^3$) | $\rho_{rel}$ (%) |
|---|---|---|---|---|---|---|
| CS | ≈68 | ≈1 | 68 | 328 | 37 | ≥99 |
| SPS | 73 | 0.93 | 67.8 | 225 | 36 | >95 |
| CSP | 73 | 0.73 | 53 | 239 | — | 92 |
| MWs | 50 | ≈0.62 | ≈31 | 445 | — | 95 |

**Table 2.**
*Magnetic parameters and relative density for top representatives of $SrFe_{12}O_{19}$ magnets manufactured following the different sintering approaches described in the present chapter, i.e., conventional sintering (CS) [61], spark plasma sintering (SPS) [18], cold sintering process (CSP) [84], and microwave sintering (MWs) [80]. Approximate values (≈) are graphically estimated from the article figures.*

alternatives available for sintering ferrites with enhanced magnetic characteristics and microstructure. To our knowledge, none of the just mentioned have yet been examined on hard hexagonal ferrites, leaving lots of room for additional study in this area.

## Acknowledgements

C.G.-M. acknowledges financial support from grant RYC2021-031181-I funded by MCIN/AEI/10.13039/501100011033 and by the "European Union NextGenerationEU/PRTR". M.S.-M. acknowledges the financial support from the Comunidad de Madrid, Spain, through an "Atracción de Talento Investigador" fellowship (2020-T2/IND-20581). H.L.A acknowledges the financial support from The Spanish Ministry of Universities (Ministerio de Universidades) and the European Union—NextGenerationEU through a Maria Zambrano—attraction of international talent fellowship grant.

## Author details

Cecilia Granados-Miralles[1*], Matilde Saura-Múzquiz[2] and Henrik L. Andersen[2]

1 Departamento de Electrocerámica, Instituto de Cerámica y Vidrio, CSIC, Madrid, Spain

2 Departamento de Física de Materiales, Universidad Complutense de Madrid, Madrid, Spain

*Address all correspondence to: c.granados.miralles@icv.csic.es

## References

[1] Coey JMD. Hard magnetic materials: A perspective. IEEE Transactions on Magnetics. 2011;**47**:4671-4681

[2] Jimenez-Villacorta F, Lewis LH. Advanced permanent magnetic materials. In: JMG E, editor. Nanomagnetism. Cheshire, UK: One Central Press; 2014. pp. 161-189

[3] Coey JMD. Permanent magnets: Plugging the gap. Scripta Materialia. 2012;**67**:524-529

[4] Permanent Magnets Market Size, Share & Trends Analysis Report by Material (Ferrite, NdFeB), by Application (Consumer Goods & Electronics, Energy), by Region, and Segment Forecasts, 2023-2030. 2023. Available from: https://www.grandviewresearch.com/industry-analysis/permanent-magnets-industry

[5] Skomski R. Nanomagnetics. Journal of Physics. Condensed Matter. 2003;**15**:R841-R896

[6] Cui J et al. Current progress and future challenges in rare-earth-free permanent magnets. Acta Materialia. 2018;**158**:118-137

[7] Igini M. Biden Fracking Ban at a Standstill Amid Global Energy Crisis. Earth.Org; 2022. Available from: https://earth.org/rare-earth-mining-has-devastated-chinas-environment

[8] Critical Materials Strategy Report. 2011. Available from: https://www.energy.gov/policy/downloads/2011-critical-materials-strategy

[9] Yao X. China is Moving Rapidly Up the Rare Earth Value Chain. Marsh McLennan Brink Newsletter; 2022. Available from: https://www.brinknews.com/china-is-moving-rapidly-up-the-rare-earth-value-chain/

[10] "This is What We Die for": Human Rights Abuses in the Democratic Republic of the Congo Power the Global Trade in Cobalt. 2016

[11] Pavel CC et al. Role of substitution in mitigating the supply pressure of rare earths in electric road transport applications. Sustainable Materials and Technologies. 2017;**12**:62-72

[12] de Julian Fernandez C et al. Progress and prospects of hard hexaferrites for permanent magnet applications. Journal of Physics D: Applied Physics. 2021;**54**:153001

[13] Granados-Miralles C, Jenuš P. On the potential of hard ferrite ceramics for permanent magnet technology-A review on sintering strategies. Journal of Physics D: Applied Physics. 2021;**54**:303001

[14] Global Permanent Magnets Market Report and Forecast 2020-2025. 2020. Available from: https://www.expertmarketresearch.com/reports/permanent-magnets-market

[15] Lewis LH, Jiménez-Villacorta F. Perspectives on permanent magnetic materials for energy conversion and power generation. Metallurgical and Materials Transactions A: Physical Metallurgy and Materials Science. 2013;**44**:2-20

[16] Park J et al. Coercivity of $SrFe_{12}O_{19}$ hexaferrite platelets near single domain size. IEEE Magnetics Letters. 2015;**6**:1-3

[17] Saura-Múzquiz M et al. Improved performance of $SrFe_{12}O_{19}$ bulk magnets

through bottom-up nanostructuring. Nanoscale. 2016;**8**:2857-2866

[18] Saura-Múzquiz M et al. Nanoengineered high-performance hexaferrite magnets by morphology-induced alignment of tailored nanoplatelets. ACS Applied Nano Materials. 2018;**1**:6938-6949

[19] Guzmán-Mínguez JC et al. Boosting the coercivity of SrFe 12 O 19 nanocrystalline powders obtained using the citrate combustion synthesis method. Journal of Physics D: Applied Physics. 2021;**54**:014002

[20] Kazin PE, Trusov LA, Zaitsev DD, Tretyakov YD, Jansen M. Formation of submicron-sized SrFe12-xAlxO19 with very high coercivity. Journal of Magnetism and Magnetic Materials. 2008;**320**:1068-1072

[21] Kershi R, Al-Asbahi S. Chemical synthesis and X-ray study of M-type hexagonal nano ferrite powders. Research Journal of Material Sciences. 2014;**2**:1-5

[22] Luo H, Rai BKK, Mishra SRR, Nguyen VVV, Liu JPP. Physical and magnetic properties of highly aluminum doped strontium ferrite nanoparticles prepared by auto-combustion route. Journal of Magnetism and Magnetic Materials. 2012;**324**:2602-2608

[23] Govea-Alcaide E et al. Structural and magnetic properties of La-doped strontium-hexaferrites ceramics obtained by spark-plasma sintering. Journal of Magnetism and Magnetic Materials. 2021;**533**:167966

[24] Jenuš P et al. Ferrite-based exchange-coupled hard-soft magnets fabricated by spark plasma sintering. Journal of the American Ceramic Society. 2016;**99**:1927-1934

[25] Soria GD et al. Uncorrelated magnetic domains in decoupled SrFe12O19/Co hard/soft bilayers. Journal of Physics D: Applied Physics. 2021;**54**:054003

[26] Pullar RC. Hexagonal ferrites: A review of the synthesis, properties and applications of hexaferrite ceramics. Progress in Materials Science. 2012;**57**:1191-1334

[27] Cullity BD, Graham CD. Introduction to Magnetic Materials. New Jersey, US: John Wiley & Sons; 2009. DOI: 10.1002/9780470386323

[28] Smit J, Wijn HPJ. Ferrites: Physical Properties of Ferrimagnetic Oxides in Relation to their Technical Applications. Eindhoven, The Netherlands: Philips' Technical Library; 1959

[29] Obradors X, Collomb A, Pernet M, Samaras D, Joubert JC. X-ray analysis of the structural and dynamic properties of $BaFe_{12}O_{19}$ hexagonal ferrite at room temperature. Journal of Solid State Chemistry. 1985;**56**:171-181

[30] Obradors X et al. Crystal structure of strontium hexaferrite $SrFe_{12}O_{19}$. Journal of Solid State Chemistry. 1988;**72**:218-224

[31] Coey JMD. Magnetism and Magnetic Materials. New York, US: Cambridge University Press; 2001. DOI: 10.1017/CBO9780511845000

[32] Chikazumi S. Physics of Ferromagnetism. New York, US: Oxford University Press; 1997

[33] Shirk BT, Buessem WR. Temperature dependence of $M_s$ and $K_1$ of $BaFe_{12}O_{19}$ and $SrFe_{12}O_{19}$ single crystals. Journal of Applied Physics. 1969;**40**:1294-1296

[34] Jahn L, Müller HG. The coercivity of hard ferrite single crystals. Physica Status Solidi. 1969;**35**:723-730

[35] Biesuz M, Grasso S, Sglavo VM. What's new in ceramics sintering? A short report on the latest trends and future prospects. Current Opinion in Solid State & Materials Science. 2020;**24**:100868

[36] Dho J, Lee EK, Park JY, Hur NH. Effects of the grain boundary on the coercivity of barium ferrite $BaFe_{12}O_{19}$. Journal of Magnetism and Magnetic Materials. 2005;**285**:164-168

[37] IMA-Ingeniería Magnética Aplicada SL. Available from: https://www.imamagnets.com/en/ferrite-magnets/

[38] Grades of Ferrite-Bunting eMagnets. Available from: https://e-magnetsuk.com/ferrite-magnets/grades-of-ferrite/

[39] Ferrite Magnets-TDK Corporation. Available from: https://product.tdk.com/info/en/catalog/datasheets/magnet_fb_summary_en.pdf

[40] Hitachi Metals L. Permanent Magnets-Hitachi. Available from: https://www.proterial.com/e/products/auto/el/p03_05.html

[41] Stuijts AL, Rathenau GW, Weber GH. Ferroxdure II and III, anisotropic permanent-magnet materials. Philips Technical Review. 1954;**16**:141-147

[42] Christensen M, Saura-Múzquiz M, Stingaciu M. A Permanent Magnetic Material. 2016. patent number: US20160167978A1, assignee: Aarhus Universitet. Available from: https://patents.google.com/patent/US20160167978A1/en?oq=US20160167978

[43] Serrano A, García E, Fernández JF, Granados-Miralles C, Quesada A. Procedimiento de obtención de un imán permanente de cerámica magnéticamente anisótropo y denso. 2021. patent number: WO2021160909A1, assignee: Consejo Superior de Investigaciones Cientificas CSIC. Available from: https://patents.google.com/patent/WO2021160909A1/es?oq=WO+2021%2f160909+A1

[44] Went JJ, Rathenau GW, Gorter EW, Van Oosterhout G. Ferroxdure, a class of new permanent magnet materials. Philips Technical Review. 1952;**13**:194

[45] Went JJ, Rathenau GW, Gorter EW, van Oosterhout GW. Hexagonal iron-oxide compounds as permanent-magnet materials. Physics Review. 1952;**86**:424-425

[46] Ghate BB, Goldman A. Ferrimagnetic Ceramics. In: Materials Science and Technology. Weinheim, DE: Wiley-VCH Verlag GmbH & Co. KGaA; 2006. DOI: 10.1002/9783527603978.mst0129

[47] Gjørup FH, Saura-Múzquiz M, Ahlburg JV, Andersen HL, Christensen M. Coercivity enhancement of strontium hexaferrite nano-crystallites through morphology controlled annealing. Materialia. 2018;**4**:203-210

[48] Chen I-W, Wang X-H. Sintering dense nanocrystalline ceramics without final-stage grain growth. Nature. 2000;**404**:168-171

[49] Groza JR. Nanosintering. Nanostructured Materials. 1999;**12**:987-992

[50] Bansal NP, Boccaccini AR. Ceramics and Composites Processing Methods. New Jersey, US: John Wiley & Sons, Inc.; 2012. DOI: 10.1002/9781118176665

[51] El Shater RE, El-Ghazzawy EH, El-Nimr MK. Study of the sintering temperature and the sintering time period effects on the structural and magnetic properties of M-type hexaferrite $BaFe_{12}O_{19}$. Journal of Alloys and Compounds. 2018;**739**:327-334

[52] Kools F. The action of a silica additive during sintering of strontium hexaferrite part I: Preparation and examination of sintered materials, the chemical action of silica, grain growth inhibition by precipitate drag. Science of Sintering. 1985;**17**:49-62

[53] Kools F. Reaction-induced grain growth impediment during sintering of strontium hexaferrite with silica addition. Solid State Ionics. 1985;**16**:251-259

[54] Beseničar S, Drofenik M. High coercivity Sr hexaferrites. Journal of Magnetism and Magnetic Materials. 1991;**101**:307-309

[55] Kobayashi Y, Hosokawa S, Oda E. Sintered Ferrite Magnet and Its Production Method. 2013. patent number: US20130285779A1, assignee: Proterial Ltd. Available from: https://patents.google.com/patent/US9401235B2/en?oq=US20130285779A1

[56] Guzmán-Mínguez JC, Vicente-Arche LM, Granados-Miralles C, Fernández JF, Quesada A. Improvement of the magnetic properties of $SrFe_{12}O_{19}$ ceramics by tailored sintering with $SiO_2$ addition. Journal of Alloys and Compounds. 2021;**860**:157890

[57] Goldman A. Modern Ferrite Technology. New York, US: Springer; 2006. DOI: 10.1007/978-0-387-29413-1

[58] Lee JW, Cho YS, Amarakoon VRW. Improved magnetic properties and growth anisotropy of chemically modified Sr ferrites. Journal of Applied Physics. 1999;**85**:5696-5698

[59] Töpfer J, Schwarzer S, Senz S, Hesse D. Influence of SiO2 and CaO additions on the microstructure and magnetic properties of sintered Sr-hexaferrite. Journal of the European Ceramic Society. 2005;**25**:1681-1688

[60] Huang C-C et al. Influence of $CaCO_3$ and $SiO_2$ additives on magnetic properties of M-type Sr ferrites. Journal of Magnetism and Magnetic Materials. 2018;**451**:288-294

[61] Du J, Zhou T, Lian L, Liu Y, Du Y. Two-step sintering of M-type strontium ferrite with high coercivity. Ceramics International. 2019;**45**:6978-6984

[62] Guzmán-Mínguez JC, Fuertes V, Granados-Miralles C, Fernández JF, Quesada A. Greener processing of SrFe12O19 ceramic permanent magnets by two-step sintering. Ceramics International. 2021;**47**:31765-31771

[63] Grasso S, Sakka Y, Maizza G. Electric current activated/assisted sintering (ECAS): A review of patents 1906-2008. Science and Technology of Advanced Materials. 2009;**10**:053001

[64] Suárez M, et al. Challenges and Opportunities for Spark Plasma Sintering: A Key Technology for a New Generation of Materials. Chapter 13 in Sintering Applications. London, UK: InTech; 2013. DOI: 10.5772/53706

[65] Obara G, Yamamoto H, Tani M, Tokita M. Magnetic properties of spark plasma sintering magnets using fine powders prepared by mechanical compounding method. Journal of Magnetism and Magnetic Materials. 2002;**239**:464-467

[66] Munir ZA, Anselmi-Tamburini U, Ohyanagi M. The effect of electric field and pressure on the synthesis and consolidation of materials: A review of the spark plasma sintering method. Journal of Materials Science. 2006;**41**:763-777

[67] Kirchner R. "FAST"-Field Assisted Sintering Technology Basics, State of the

Art and Future Aspects. International Spring School on Field Assisted Sintering Technique; Darmstadt, DE; 2011

[68] Mazaleyrat F, Pasko A, Bartok A, LoBue M. Giant coercivity of dense nanostructured spark plasma sintered barium hexaferrite. Journal of Applied Physics. 2011;**109**:07A708

[69] Ovtar S, Le Gallet S, Minier L, Millot N, Lisjak D. Control of barium ferrite decomposition during spark plasma sintering: Towards nanostructured samples with anisotropic magnetic properties. Journal of the European Ceramic Society. 2014;**34**:337-346

[70] Stingaciu M, Topole M, McGuiness P, Christensen M. Magnetic properties of ball-milled $SrFe_{12}O_{19}$ particles consolidated by spark-plasma sintering. Scientific Reports. 2015;**5**:14112

[71] Granados-Miralles C et al. Unraveling structural and magnetic information during growth of nanocrystalline $SrFe_{12}O_{19}$. Journal of Materials Chemistry C. 2016;**4**:10903-10913

[72] Eikeland AZ, Stingaciu M, Mamakhel AH, Saura-Múzquiz M, Christensen M. Enhancement of magnetic properties through morphology control of $SrFe_{12}O_{19}$ nanocrystallites. Scientific Reports. 2018;**8**:7325

[73] Stingaciu M, Eikeland AZ, Gjørup FH, Deledda S, Christensen M. Optimization of magnetic properties in fast consolidated $SrFe_{12}O_{19}$ nanocrystallites. RSC Advances. 2019;**9**:12968-12976

[74] Saura-Múzquiz M et al. Elucidating the relationship between nanoparticle morphology, nuclear/magnetic texture and magnetic performance of sintered SrFe12O19 magnets. Nanoscale. 2020;**12**:9481-9494

[75] Vijayan H, Knudsen CG, Mørch MI, Christensen M. Ultrathin nanoplatelets of six-line ferrihydrite enhances the magnetic properties of hexaferrite. Materials Chemistry Frontiers. 2021;**5**:3699-3709

[76] Thomas-Hunt J et al. Alignment of strontium hexaferrite, by cold compaction of anisotropic non-magnetically interacting crystallites. Dalton Transactions. 2022;**51**:3884-3893

[77] Vijayan H et al. High-performance hexaferrite ceramic magnets made from nanoplatelets of Ferrihydrite by high-temperature calcination for permanent magnet applications. ACS Applied Nano Materials. 2023;**6**:8156-8167

[78] Binner JGP, Caro I, Firkins J. Microwave sintering of nanometer and micrometer ferrite powders. The Journal of Microwave Power and Electromagnetic Energy. 1999;**34**:131-136

[79] Yang Q, Zhang H, Liu Y, Wen Q. Microstructure and magnetic properties of microwave sintered M-type barium ferrite for application in LTCC devices. Materials Letters. 2009;**63**:406-408

[80] Kanagesan S et al. Effect of microwave sintering on microstructural and magnetic properties of strontium hexaferrite using sol-gel technique. Journal of Materials Science: Materials in Electronics. 2013;**24**:3881-3884

[81] Guo J et al. Cold sintering: A paradigm shift for processing and integration of ceramics. Angewandte Chemie International Edition. 2016;**55**:11457-11461

[82] Serrano A et al. Hexaferrite-based permanent magnets with upper magnetic properties by cold sintering process via a non-aqueous solvent. Acta Materialia. 2021;**219**:117262

[83] Serrano A et al. Effect of organic solvent on the cold sintering processing of SrFe12O19 platelet-based permanent magnets. Journal of the European Ceramic Society. 2022;**42**:1014-1022

[84] García-Martín E et al. Dense strontium hexaferrite-based permanent magnet composites assisted by cold sintering process. Journal of Alloys and Compounds. 2022;**917**:165531

[85] Grasso S et al. A review of cold sintering processes. Advances in Applied Ceramics. 2020;**119**:115-143

[86] Yu M, Grasso S, Mckinnon R, Saunders T, Reece MJ. Review of flash sintering: Materials, mechanisms and modelling. Advances in Applied Ceramics. 2017;**116**:24-60

[87] Biesuz M, Sglavo VM. Flash sintering of ceramics. Journal of the European Ceramic Society. 2019;**39**:115-143

[88] Grasso S et al. Flash spark plasma sintering (FSPS) of α and β SiC. Journal of the American Ceramic Society. 2016;**99**:1534-1543

[89] Muche DNF, Drazin JW, Mardinly J, Dey S, Castro RHR. Colossal grain boundary strengthening in ultrafine nanocrystalline oxides. Materials Letters. 2017;**186**:298-300

[90] Herisson de Beauvoir T, Sangregorio A, Cornu I, Elissalde C, Josse M. Cool-SPS: An opportunity for low temperature sintering of thermodynamically fragile materials. Journal of Materials Chemistry C. 2018;**6**:2229-2233

Chapter 5

# On Recent Progress on Core Shell Nanostructures of Colossal Permittivity Materials for Capacitors: Synthesis and Dielectric Properties

*Cécile Autret-Lambert, Samir Merad, Sonia De Almeida-Didry, Olivier Motret and François Gervais*

## Abstract

Dielectric materials with colossal permittivity show promise for the development and miniaturization of high-performance capacitors. $CaCu_3Ti_4O_{12}$ (CCTO) improvement for multilayer ceramic capacitors (MLCCs) has been achieved. CCTO shows a large $\varepsilon'$ of $\sim 10^4$ over a temperature range. This behavior is due to a potential barrier at the grain boundaries (GBs). CCTO ceramics have an electrically heterogeneous microstructure with semi-conducting grains and more insulating GBs, analyzed by an internal barrier layer capacitor (IBLC) structure model. Therefore, the dielectric properties of these materials can be improved by changing the electrical properties of the grains and GBs. In this context, core-shell approaches to control the GBs have been developed. This chapter presents advanced synthesis techniques (by chemistry way but also by cold plasma) to design the dielectric grains of CCTO by shells of different nature, morphology and crystallinity and shows the impact on the macroscopic properties.

**Keywords:** core-shell particles, CCTO, dielectric properties, IBLC, pulsed cold plasma

## 1. Introduction

Mobile electronic devices connected to the broadband network and with high transmission capacity are widespread. The trend today is toward miniaturization, better quality and lower power consumption. Capacitors, filters, resistors are among the passive components widely used. Conventional capacitors, i.e. monolayer capacitors have been primarily used. However, the MLCC component has taken the lead due to its high capacity properties even with smaller sizes, high reliability and excellent frequency characteristics [1–5]. Recently, the worldwide shortage of electronic components such as MLCC capacitors has become a major problem. In recent years, electric vehicles, which require a large number of capacitors because of their electronic control and automation systems, have a great need of them. Many dielectric layers that are stacked alternately by internal electrodes connected in parallel constitute the MLCC device. This configuration allows increasing the capacitance of the components [6, 7]. However, the capacitance can be enhanced with thin dielectric layers and with the number of stacked layers but also by using high-k dielectrics. The $BaTiO_3$-based dielectric material is used as the reference material in the multilayer ceramic capacitor (MLCC) industry with noble metals such as palladium or nickel as internal electrode. MLCCs suffer from various disadvantages. Due to the different materials (dielectric material and electrodes) with different sintering temperatures and coefficients of expansion, cracking or delamination occurs at the interfaces of the layers. This residual stress has an unfavorable effect on the mechanical and electrical properties. To overcome these problems, another promising solution has been developed, the ceramic-metal composite materials (CMC), in which metal particles are dispersed in the ceramic matrix without interacting with the network structure. In these materials, ceramic acts as an insulating phase while metal particles are used as conductors. The enhancement of the permittivity in these systems (MLCC and CMC) is due to the ceramic/metal interface [8, 9].

However, modern technologies require new materials to develop new technologies. Indeed, the nanotechnology industry continues to grow rapidly in order to expand the field of applications but also to improve the devices used today. Thus two ways are possible, the fabrication of nanomaterials or so-called "core-shell" materials. With regard to nanomaterials, a great effort has been made toward the synthesis and application of nanoparticles because of their unusual physical and chemical properties due to their very different specific surface compared to the bulk properties [10, 11].

More recently, research has focused on nano-structured "core-shell" nanoparticles to adapt structural, magnetic, electrical properties but also to change surface properties. The purpose of the "core-shell" is to coat particles with another type of material. The presence of the shell can modify the surface, to increase the core material functionality or stability. For example, the temperature stability of the dielectric properties in the capacitor is always asked to extend the field of applications. One possibility is to use these core-shell nanostructures which have a different chemical composition between bulk grain and grain boundary (GB) [12, 13]. In the literature, many different materials such as metal oxides, noble metals or polymers are used as shells to coat the $BaTiO_3$ particles [14, 15]. In the field of high permittivity dielectrics, such as $BaTiO_3$, it is interesting to coat with a stable insulating material such as $SiO_2$, which also has the advantage of being stable, chemically inert and non-toxic [16, 17]. Moreover, the preparation of the $SiO_2$ shell is well-known. And a homogeneous distribution of $SiO_2$ layer coating is very important to prevent the exaggerated grain growth and the segregation of $SiO_2$ which can alter the dielectric properties [18, 19]. In 2008, Wei et al. proposed a method, in their patent, for elaborating core@shell structured dielectric particles consisting of a conductor core and one or more layers of thinly-coated dielectric material for use in multilayer ceramic capacitors (MLCC) applications [20]. This new configuration was considered to replace conventional $BaTiO_3$ dielectric materials as capacitor active layers.

From a general point, the permittivity results from both intrinsic or extrinsic properties, the latter been associated with possible existence of domains, defects, doping, internal strains… [21–23]. In some cases, a strong increase in permittivity, slightly constant on a large range of temperatures can be observed depending on the doped $BaTiO_3$ substitution [24] or on the synthesis atmosphere which can create vacancies or doping on the grain surface but not penetrate the interior of the grains [25]. Permittivity values as high as $10^5$ have been reported for nanostructured materials, such as core-shell structures with a $BaTiO_3$ core and a shell made of another material like $SiO_2$, $Al_2O_3$... [26]. These good properties are used in industrial capacitors of type III or X7R4-5, where high permittivity and capacitance are needed. However, these capacitors come with the drawback of low breakdown voltage, high frequency sensitivity and temperature-related performance variation. For example, the high permittivity of $BaTiO_3$ originates from FE-PE phase transition and it means that the properties vary strongly with temperature. In this work, we will focus on the perovskite-like oxide $CaCu_3Ti_4O_{12}$ (CCTO) proved to be one of the best candidates due to its interesting dielectric constant $\varepsilon r \approx 10^4$–$10^5$ in a large temperature and frequency range [27, 28]. Despite its colossal dielectric constant, CCTO material exhibits a strong loss tangent and a low breakdown voltage that limits its use in industrial application [29]. On the other hand, numerous studies have been conducted and many mechanisms have been proposed to explain the origin of the colossal dielectric constant [21, 30]. The most widely accepted mechanism, known as the Internal Barrier Layer capacitance (IBLC), consists of semi-conducting grains and insulating GB. The grains would behave as the electrodes of a capacitor and GBs act like the dielectric. This configuration consists of many capacitors in series that could explain the high dielectric constant. Some authors attempt to approximate the permittivity as a function of the grain boundary permittivity using this formula:

$$\varepsilon = \varepsilon_{gb}\, A/t \tag{1}$$

where $\varepsilon_{gb}$ is the permittivity of the grain boundary, $A$ is the size of the grain (μm) and $t$ is the thickness of the grain boundary (μm) [31–33].

Thus the idea of the existence of semi-conductor grains and insulating GBs leads us to the idea of making this configuration by developing nanostructured "core-shell" materials. For that, we envisaged to optimize the synthesis methods for the production of CCTO-based core-shell nanostructures. The main problem when coating the high-k material (CCTO) is achieving control over homogeneity and the shell thickness. This control is crucial not only for regulating the properties of the GBs but also for limiting the diffusion of chemical elements between the core and the shell, which can potentially alter the material's properties. Through an overview of recent results, we will highlight the advantages and disadvantages of synthesis for different coatings. We will also show a new way of synthesizing "core-shell" which is promising for shell quality.

## 2. $CaCu_3Ti_4O_{12}$ particles

### 2.1 Structure

The crystal structure of $CaCu_3Ti_4O_{12}$ was discovered at the first time by Bochu et al. and confirmed by Subramanian et al. [30, 34]. CCTO crystallizes in a cubic structure that is derived from the ideal cubic perovskite structure ($ABO_3$). In this structure, there is an arrangement of $Cu^{2+}$ and $Ca^{2+}$ cations constituting the A-site, resulting in the chemical formula $AA'_3B_4O_{12}$ formula. The Jahn-Teller effect causes the $Cu^{2+}$ cation site to occupy a square-planar environment instead of a 12-coordination site as the ideal perovskite, which induces a strong tilting of $TiO_6$ octahedra.

The unit-cell body-centered cubic CCTO with the Im-3 space group consists of two formula units with cell parameters a ≈ 7.39 Å. The structural analysis as a function of temperature showed no structural transition, indicating the absence of a ferroelectric transition [35].

The dielectric permittivity of CCTO has been reported in a large range from 100 to $10^5$ in films, single crystals or bulk [30, 36–38] and in certain cases, the permittivity remains constant in a large domain of temperature. Even if today an explanation is not convincing, the community agrees that there is no ferroelectric transition and that the behavior is not intrinsic [39]. Then it was assumed that the colossal permittivity could find its origin from extrinsic properties of the material. In 2002, Sinclair et al. have proposed a theory to explain this behavior, based on their impedance spectroscopy results [40]. Indeed, by association of semi-conducting grains and insulating grain-boundaries, there would be an internal barriers phenomenon named "Internal Barrier Layer Capacitance" effect (IBLC). However, the strong ε value observed in single crystals indicates that the insulating barriers exist within the crystals themselves rather than between them, indicating the presence of twin boundaries that act like barriers [30]. Even if the structure has a decisive role, it seems that the good properties are enhanced by the microstructure of the compounds and specifically by the GBs.

## 2.2 Parameters influencing the CCTO properties

The CCTO dielectric properties may be affected by many factors, including the synthesis method, sintering temperature and time, doping ions and electrode materials [36, 41–44].

### *2.2.1 Sintering temperature and time effect*

Brize et al. worked on the effects of the synthesis conditions by using various conditions such as the powder mixing, the time, the calcination temperature [45]. Although the development of oxide materials by solid-state reaction remains a widely used method of synthesis, they preferred to use the modified citrate process [46]. Nanometric powder can be obtained with this synthesis method depending on the time and the calcination temperature. They choose the temperature from 500°C up to 750°C and the time from 8 h up to 20. All pellets were sintered at the same temperature, which was maintained at 1050°C for 24 h. They showed that the best properties were observed for the powder calcined at 500°C for 20 h with dielectric constant close to $10^5$. This strong dielectric constant was related to the large grain size and density. The grain size effect was already reported by Bender et al. as early as 2005 [47]. Later, Sonia et al. confirmed that the grain size favors high permittivity (**Figure 1**) [48]. These observations validate the proportionality of $\varepsilon_r$ with the size A of the grains of the IBLC model (Eq. (1)).

### *2.2.2 Doping effect*

The cationic substitution on the A, A' and B sites can change the dielectric and electric properties. Numerous papers in the literature relate the CCTO doping by one element and more recently by two elements, i.e. the co-doping [49]. Clearly, the substitution and co-substitution have a strong influence on dielectric ($\varepsilon_r$ and tan δ) and electric properties as summarized by Ahmadipour et al. [50] and S. De Almeida-Didry et al. [51]. The list is not exhaustive, but taking into account the results, the best

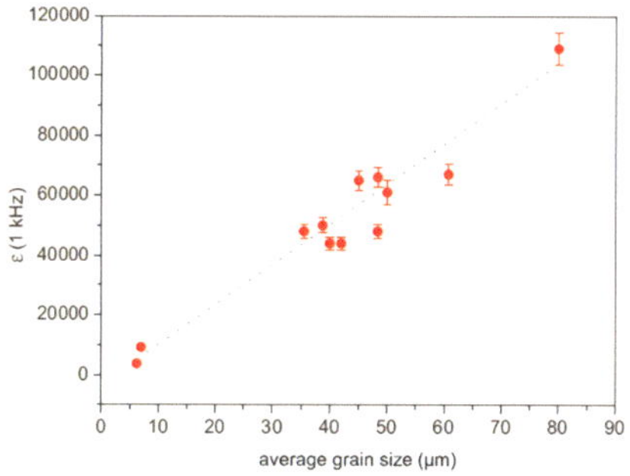


**Figure 1.**
*Increase of permittivity of undoped CCTO with the average size of the grains [1–5].*

dielectric constant was given for the CCTO containing less copper but at the expense of the loss tangent. It was also demonstrated that the colossal dielectric properties were collapsed by Mn and Fe doping even with a slight substitution equal to 0.5% or 1% [38, 52–54]. The permittivity for further materials with doping or co-doping is between these two extremes. However, none of them satisfies the properties required for use in the industry. To understand the mechanism, Sinclair et al. did some complex impedance spectroscopy measurements [40]. They showed that the dielectric properties originate from the electric grain behavior and GBs. Indeed, the grains have a semi-conducting character and the grains boundaries are insulating. This difference between insulating GBs and semi-conducting bulk can be schematized by IBLC structure, which effectively explains the properties of these ceramics. Numerous studies have shown that copper comes out of CCTO grains during the sintering process and this causes a difference in electrical behavior in grains and GBs. Grains have copper vacancies and become conductive while GBs are insulating, leading to the formation of the IBLC structure [55–59]. Concerning the Mn and Fe doping, Krohns et al. put forward the reduction of grain conductivity that implies that the phenomenal IBLC is suppressed. They argued that the permittivity is related to the polar character of phonons without amplification by barrier layers [38]. Finally, the cations, which seem to be good candidates, are $Cr^{3+}$, $La^{3+}$, $Al^{3+}$, however, resistivity is still too far from the values required for industrialization [49, 60, 61].

Sinclair et al. have shown the existence of Schottky barriers at the interfaces, and they insist on the role of the GBs between semi-conducting grains. The Schottky barriers likely are at the origin of the non-linear behavior of J-E characteristics, where E is the electric field and J the current density that is a drawback in a charge storage device. Sinclair et al. have concluded their paper by the fact that CCTO ceramics with grain size of few micrometers do not present this non-linearity and could be good candidates for applications due to their permittivity resulting from a "core-shell" structure of insulating GBs and conducting grains [40].

### 2.3 The nature of grain boundaries understanding

To summarize, the studies conducted in the literature revealed that the GBs play a key role since the relative permittivity could be estimated according to the Eq. (1). In this purpose, the precise study of the structure, grain boundaries composition could be crucial. From a sintered ceramic of CCTO with large grains and small GBs, a cross-section sample was prepared using a dual-beam focused ion beam (FIB) based on a focused $Ga^+$ ion beam and transferred to a Cu-grid using a micromanipulator. The

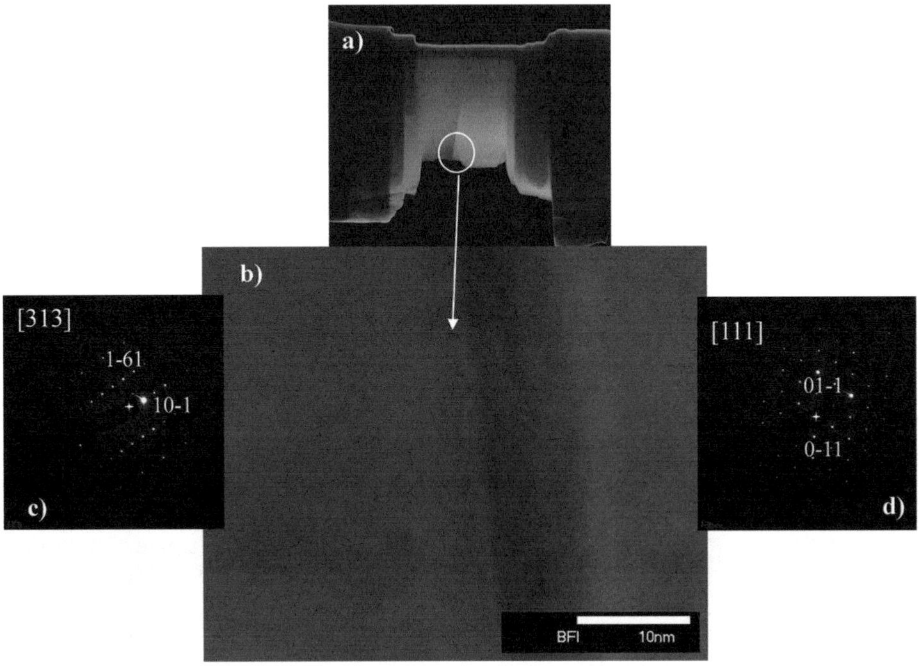


**Figure 2.**
*(a) TEM lamella preparation by FIB, (b) STEM images showing the two grains separated by a dark line (grain boundary) and (c) and (d) SAED patterns of orientated grains.*

lamella was located on a GB surrounded by two grains. **Figure 2** illustrates a SEM image of the TEM lamella with two grains separated by a GB, which appears with a darker line. SAED studies permit to obtain ED patterns for each well-crystallized grain. There is no correlation between the orientations of the grains, the main dots are indexed and the zone axes are different (**Figure 2**).

The HRTEM image (**Figure 3**) shows a view of the crystal with a uniform background contrast consisting of a very regular array of bright dots spaced by about 5.2 Å for the <011>. No defect is evidenced. The GB cannot be aligned; its width is of a few nanometers about 5 nm (insert **Figure 3**). The observed contrast is confused and could correspond to amorphous phase or to the superimposition of two crystallized domains. HRTEM cannot bring any more information on the GB, so elementary cartography by STEM was done allowing analysis at nanometric scale. EDX line profile was performed across both grains. No change of the element ratio was observed at the interface. This study at nanometric scale was also studied by B. Domengès et al. on two different lamellar. They showed an identical result to ours on one cross-section and the other showed the existence of Cu-rich phase that can answer the question of what type of phase is constituted GB explaining the IBLC mechanism [62].

## 3. Strategies of design of core-shell CCTO@MO

### 3.1 Synthesis of the core (CCTO)

Solid-state reaction is the predominant method used to prepare the core composed of $CaCu_3Ti_4O_{12}$. This method is simple and allows the preparation of the materials

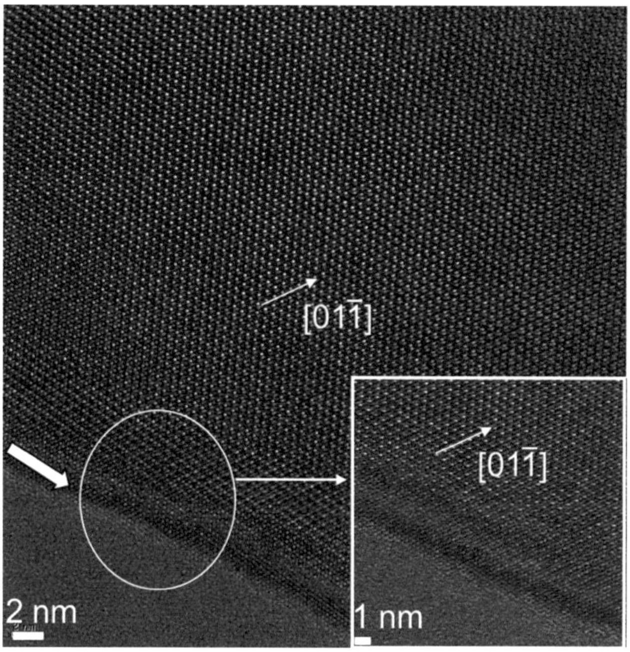


**Figure 3.**
*High resolution TEM images taken with [111] zone axis. In inset, zoom on the GB.*

from a mixture of solid starting materials. The reagents $CaCO_3$, CuO and $TiO_2$ were weighted in stoichiometric proportions. The precursors are then mixed and ground in isopropanol using zirconia balls for 6 h to homogenize the powder. Finally, the resulting powder is put in a muffle oven for a heat treatment of 950°C for 24 h. This synthesis process was chosen for the reproducibility of the structure and microstructure of the powder, more particularly, the size of the grains remains constant and close to 1 μm and the IBLC mechanism requires large grains (Eq. (1)). The quality of the powder was checked by XRD diffraction and the average size of the grains was analyzed using laser granulometry and SEM (**Figure 4**).

### 3.2 Surface charge of the CCTO particles

The dispersion of CCTO powder is a key point for the coating. This is very critical for nanopowders with size particles inferior to 100 nm. This is not the case of the CCTO used in this study, since the particle size is close to 1 μm. However, it is very important to be able to disperse homogeneously the particles in a solvent. For this, the particle surface charge needs to be known in an aqueous or non-aqueous medium. The particle surface properties are related to the potential of the particles in different environment. The surface charge is a property that all materials possess or acquire when they are suspended in a solution. The charge was evaluated by zeta potential measurements in relation to pH value. The started solutions contain CCTO sample mixed in a KOH solution, which is titrated with an HCl solution to decrease pH until 2. As the solution is titrated versus pH, the intercept is called iso-electric point (IEP) and corresponds at the maximum of instability of the particle dispersion. On the other hand, the maximum stability is obtained at high absolute potential values. So taking into account the **Figure 5**, the stability of the CCTO is found for a basic pH.

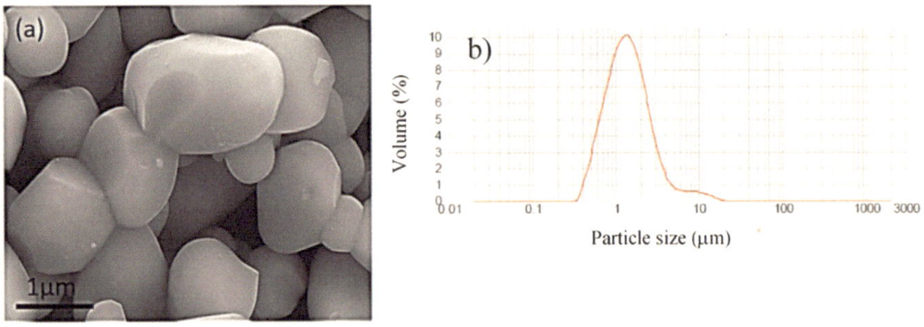


**Figure 4.**
*(a) SEM images of CCTO particles synthesized by solid state reaction and (b) particle size distribution by granulometry.*

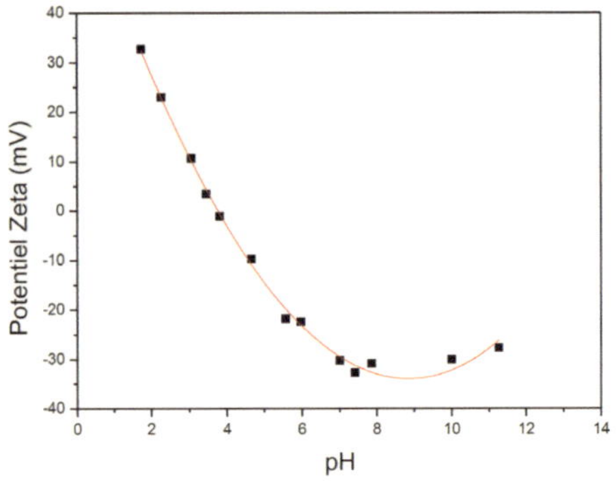


**Figure 5.**
*Zeta potential of CCTO versus pH.*

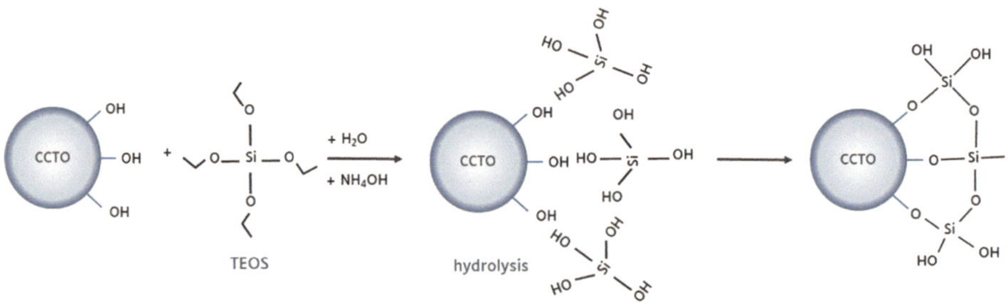


**Figure 6.**
*Mechanism of silica shell formation on the CCTO particles.*

## 3.3 Silica-based core-shell CCTO

The schematic illustration of the core-shell structured process is depicted in **Figure 6**. The method is based on the Stöber process [63] in which silica shell is formed around a core via the hydrolysis followed by a condensation reaction of a silicate precursor—in this work, we used TEOS. The Stöber method takes place at a basic pH close to 10. The negative zeta potential of CCTO (**Figure 5**) means that OH groups are on the surface, and then the silicate group from the TEOS precursor attaches to the -OH groups. After this first layer, additional layers are formed by hydrolysis and condensation of TEOS.

The process of CCTO core@silica-shell was as follows. At first, CCTO powder was added in a mixed solution of ethanol/$H_2O$/ammonia (75/23.5/1.5%vol) and the solutions were adjusted to a pH value of 10. Several concentrations of CCTO in the solution were tested (2 g/l, 5 g/l, 10 g/l). No influence on the coating and dielectric properties was identified. For industrial applications, the powder amount must be important, so the concentration of 10 g/L was chosen. The solution was stirred at room temperature during 1 h. Afterwards, tetraethyl orthosilicate (TEOS) was slowly added in the first solution under magnetic stirring. The final solution was mixed for 12 h. Finally, the mixed solutions were centrifuged and washed with ethanol several times and then the powders were dried at 80°C for 12 h to obtain core-shell structured CCTO@$SiO_2$ particles. This process and the volume of ammonium make it possible to obtain a homogeneous and uniform $SiO_2$ shell surrounding the CCTO particles [64]. The thickness of the silica shell can be calculated by changing the amount of TEOS used in the following reaction:

$$V_{TEOS} = \frac{M_{TEOS} \times \rho_{SiO_2} \times \pi}{\rho_{TEOS} \times M_{SiO_2} \times 6} \times Np(CCTO)\left(D^3 - d^3\right) \quad (2)$$

with

$$N_P = \frac{m_{(CCTO)} \times 6}{\left(\rho_{(CCTO)} \pi d^3\right)} \quad (3)$$

Where (d) is the CCTO particle diameter (μm), (D) is the core-shell particle diameter (μm). The CCTO particles are assumed to be spherical with 1 μm as diameter (d), as observed on the as-synthesized CCTO particles. The eq. (2) is as a function of the molecular weights of $SiO_2$ formula unit and TEOS (g.mol$^{-1}$), the densities of silica and TEOS (2 g.cm$^{-3}$ and 0.934 g.cm$^{-3}$, respectively). The amount of TEOS was calculated from the average size CCTO particles (~1 μm) to obtain a silica nanocoating from 1 to 125 nm.

### 3.4 Shell thickness and uniformity

To confirm the shell, transmission electronic microscopy was used. HR-TEM images demonstrated that the particles have a core-shell structure with different contrast for both phases, light contrast for the silica shell and dark contrast for the core of CCTO (**Figure 7**). To confirm these contrasts, STEM-EDS mapping of CCTO@$SiO_2$ particles was conducted for a thickness close to 10 nm (**Figure 8a**). The CCTO particles were successfully coated by silica shell. As shown in **Figure 7**, we found that the shells were uniform and homogeneous. The experimental shell thickness was found to be consistent with the calculated thickness, at least up to 20 nm. Thereafter, it is observed that the thickness remains constant for a thickness slightly above 20 nm (**Figure 8c**). Thus, for a high concentration of TEOS, the appearance of spherical $SiO_2$ particles is observed (**Figure 8** (a thickness of 65 nm image)). The formation of spherical $SiO_2$ particles in addition to the particles coating is always a side effect in the coating process. Therefore, it is crucial to control the proportion not only of TEOS but also of ammonium that serves as a catalyst. In the literature, it has been shown that for an optimized ammonium concentration, there is a minimum concentration of TEOS needed to guarantee full coating of the particles, and then the shell thickness can be adjusted with the concentration of TEOS [64–66].

**Figure 7.**
*TEM images of $CCTO@SiO_2$ for thickness shell from 1 nm up to 65 nm. The light contrast corresponds to the amorphous shell.*

## 3.5 Surfactants effects on core-shell particles

The coating by chemical process to obtain a homogeneous shell is sometimes difficult due to the nature of surface (charge, hydrophobic, ... ) and mutual chemical interactions. To improve the shell, several agents can functionalize the core particles. Moreover, the surface modification is demonstrated to improve the dispersibility of particles in a solvent. Researchers have also demonstrated that the surface modification has an influence on grain growth [67, 68]. PEG, Brj58, F127 agents are used as in-situ surfactants. CCTO particles with the modified surface were then used in the Stöber process to coat them with a silica shell. Jana et al. described different mechanisms for the functionalization of core particles with the Stöber method [69]. Brij58/F127 agents are surfactants characterized by CMC (Critical Micellar Concentration). CMC is the concentration of surfactant in an environment from which micelles form spontaneously. Below this, the surfactant forms a layer on the surface of the liquid and the rest is dispersed in the solution. We tested the concentration below and above the CMC of each surfactant. Surfactants have a crucial role in improving stability of colloidal system. To investigate the stability of these systems, zeta potentials were measured at pH = 10 (**Figure 9**).

As shown in **Figure 9**, the effect of surfactant on CCTO particles is not the one expected for a pH equal to 10, which is the value of solutions in the Stöber process. However, regardless of the surfactant used, coating up to 20 nm (by step of 5 nm) was possible. For each system, the pellets from materials coated with a 20 nm thick $SiO_2$ shell, were sintered at 1050°C for 24 h and the dielectric properties were

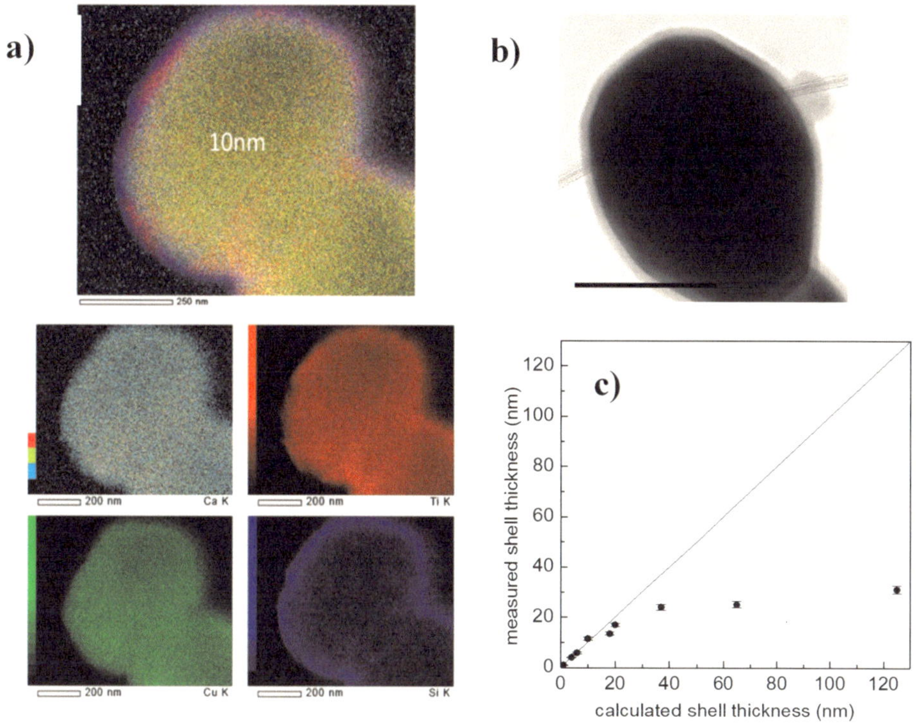


**Figure 8.**
*(a) Compositional elemental mapping of a core-shell nanoparticle using STEM-EDX, (b) bright-field transmission electron microscopy (TEM) image of a core-shell nanoparticle and (c) real shell thickness for all samples as function as shell thickness calculated with Eq. (2).*

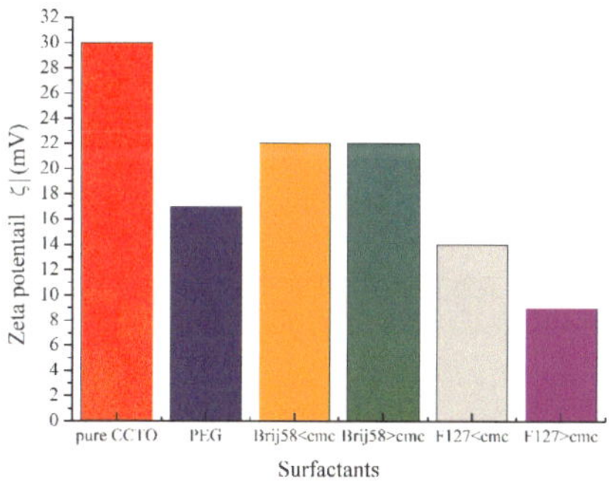


**Figure 9.**
*Zeta potential measured for CCTO without and with surfactants (PEG, Brij and F127).*

measured (**Figure 10**). The good properties obtained in the table are linked with the high average grain size. The benefit of the surfactant in the case of $SiO_2$ shell is unclear and their exact role in preventing agglomeration is still under discussion. This fact was already noted in the work of Hai et al. on $BaTiO_3$ particles [70]. However, for other shells like $Al_2O_3$, the surfactant helped us to coat uniformly the particles.

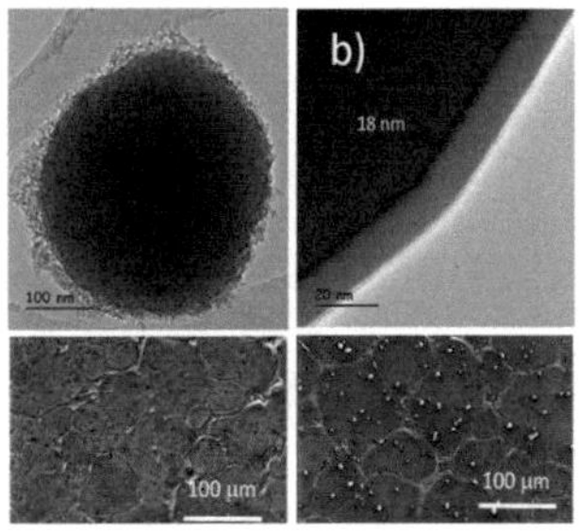


c)

| | ε (1 kHz) | Tan δ | ρ (MΩ.cm) |
|---|---|---|---|
| CCTO pur | 30000 | 0.1-0.12 | 130 |
| PEG | 50000 | 0.1-0.12 | 30 |
| Brij 58 (non ionic) | 40000 | 0.1-0.12 | 50 |
| Pluronic F127 (Amphiphile) | 50000 | 0.1-0.12 | 100 |

**Figure 10.**
*(a) CCTO@SiO$_2$ (20 nm) with PEG (b) CCTO@SiO$_2$ (20 nm) with Brij58 and below for both SEM image of the corresponding pellets showing the grain and GBs. (c) table: dielectric properties for each pellets of CCTO without and with surfactant coated by 20 nm.*

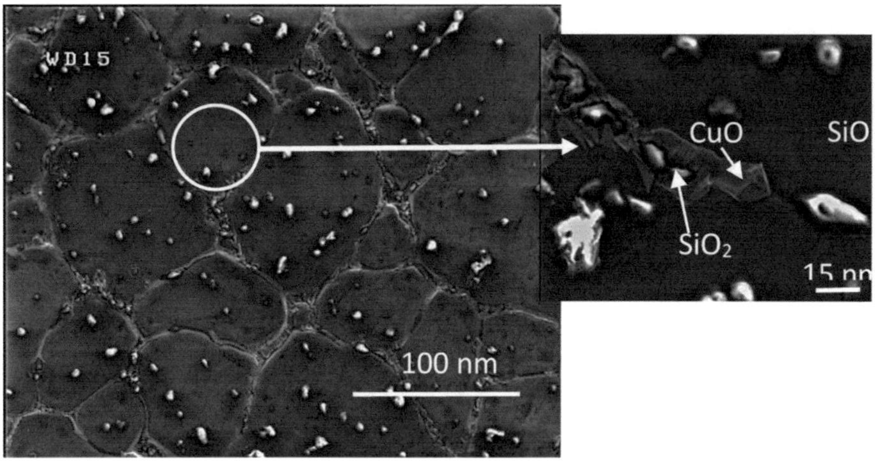


**Figure 11.**
*SEM image of the microstructure showing the grain and GBs with some SiO$_2$ particles on the grains. In enlarged zone, the GBs showing the presence of CuO and SiO$_2$ phases.*

## 3.6 The thickness shell effect on the properties

In order to investigate the impact of silica shell on the dielectric properties, pellets were made at 1050°C for 24 h. The average grain size of CCTO@SiO$_2$, about 80 µm is similar to un-doped CCTO. For this one, previous reports have shown that during sintering, a liquid copper phase appeared and led to an enhancement of the grain size [40, 71]. It seems that the same mechanisms are responsible for the grains growth of SiO$_2$-coated samples and that this dopant does not play a role in growth (**Figure 11**). It was also noted that the ceramic density is between 91 and 96% and no correlation with SiO$_2$ is observed. In addition, the initial core-shell structure is lost after thermal treatment. However, as shown in **Figure 11**, it can be noted that SiO$_2$ grains are located on the grains and also between the grains, i.e. constituting the GBs. CuO existence at the GBs also occurs, as attempted in the CCTO ceramics.

Dielectric measurements on the core-shell materials indicated that the dielectric constant ($\varepsilon_r$) slightly decreased with shell thickness values less than 20 nm, but then the drop is drastic and follows an exponential function (**Figure 12**). This fact could be

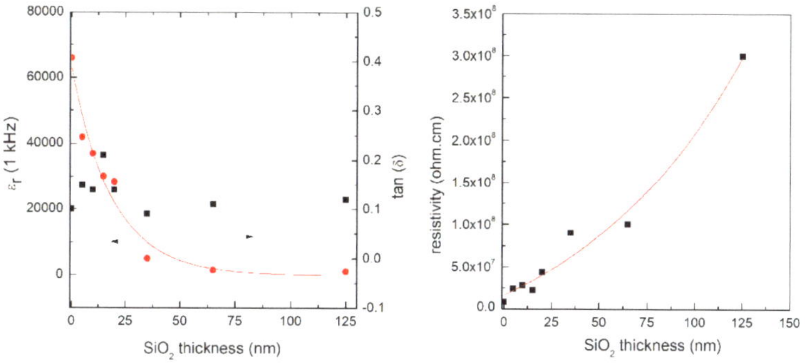


**Figure 12.**
*At left, evolution of dielectric constant and loss tangent and at right resistivity as function as $SiO_2$ shell thickness.*

explained by the appearance during the synthesis of $SiO_2$ spherical particles, the higher the TEOS concentration is, the higher the amount of $SiO_2$ is (for example, the TEM image of particles coated with 65 nm in **Figure 7**). When the samples are sintered at high temperature, $SiO_2$ particles crystallize into $SiO_2$ quartz and cristobalite. $SiO_2$ has a very low dielectric constant and, therefore, reduces $\varepsilon_r$ of the ceramic. Nevertheless, the presence of the $SiO_2$ phase at the GB limits the loss tangents, which remain around 0.1 regardless of the composition. Moreover, in accordance with the drop of the dielectric constant, the resistivity increases to very high values.

Although the $SiO_2$ shell does not improve the dielectric properties as desired, the effect is positive. Control of the GB thickness is possible with an accuracy of 1 nm and the properties for shell values <20 nm are slightly better. Beyond the amorphous silica coating, it was interesting to investigate other insulating shells that might exhibit better properties. In the literature, significant effects of the $Al_2O_3$ and $TiO_2$ shells on the resulting dielectric behaviors of the ceramics were observed. A large dielectric constant was obtained for $BaTiO_3$@$TiO_2$ due to the interfaces created by the introduction of the intermediate $TiO_2$ layer [72]. $BaTiO_3$@$Al_2O_3$ showed an enhanced dielectric breakdown strength and decreased energy loss without degradation of the dielectric constant [73].

### 3.7 Aluminum oxide-based core-shell nanocomposites

The process used to synthesize CCTO@$Al_2O_3$ core-shell particles is slightly different from the one used for $SiO_2$ coating, but it is derived from Stöber and Peng methods [63, 74]. It is well-known that the growth of a uniform $Al_2O_3$ shell outside the surface is very challenging due to the large lattice mismatch between the core and $Al_2O_3$, along with their relatively weak chemical interaction [75]. Aluminum propoxide was dissolved in an ethanol solution, and the precursor amount was calculated using the same Eq. (2) as was used for $SiO_2$, based on the calculated average size of the CCTO particles. CCTO powder was identical for all coated samples, and as a result, the average grain size remained at 1 μm. After the dissolution of the precursor, a CCTO powder mixture with a concentration of 10 g/l with ethanol was dropped into the suspension, and then dispersed by sonification for 20 min. In parallel, a 1:5 (v/v) mixture of water and ethanol (50 mL) was added slowly drop by drop, and the mixture reaction was ultrasonicated for 2 h. The particles were collected by

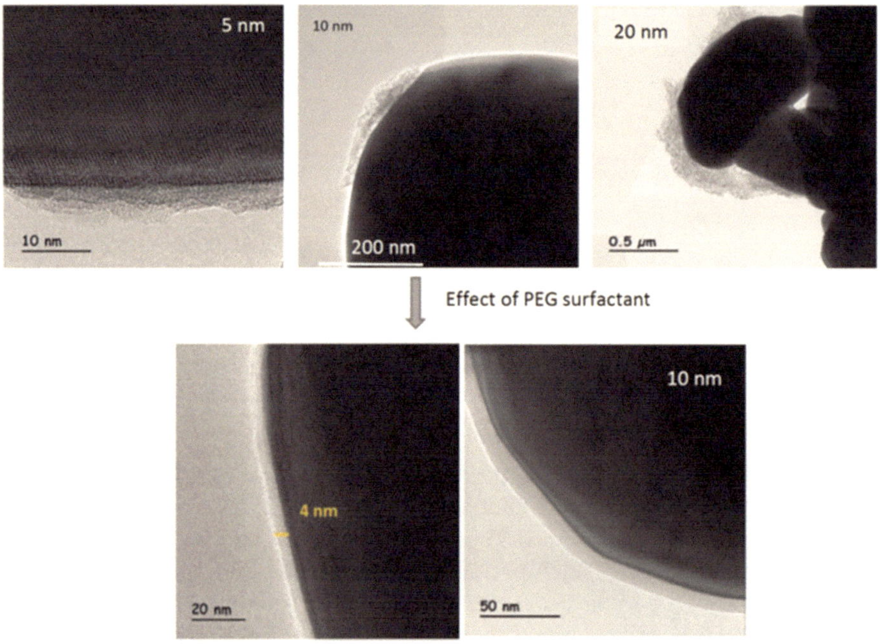


**Figure 13.**
*TEM images of CCTO@$Al_2O_3$ for thickness shell from 1 nm up to 20 nm, at the top the CCTO is not functionalized, at the bottom the CCTO is worked with PEG. The light contrast corresponds to the amorphous shell.*

centrifugation and washed with ethanol and ultrapure water. From this power, pellets were pressed and sintered at 1050°C during 24 h.

**Figure 13** depicts the TEM images of CCT@$Al_2O_3$ for several thicknesses. These images clearly show that the coating is not homogeneous and uniform. Its thickness varies along the grain surface, and sometimes no coating is visible at all. This observation led us to modify the solution containing the CCTO particles. To avoid particle agglomeration, the surfactant polyethylene glycol (PEG) was used, which has been reported to be the efficient protecting agents to stabilize a solution. A series of samples were realized with CCTO functionalized by PEG and subsequently coated with aluminum propoxide using the same process (precursors, proportion of solvent, time sintered temperature ... ). The TEM images confirmed a better coating for all the samples with a clear core-shell nanostructure of CCTO@$Al_2O_3$ with homogeneous and compact shell layer. CCTO grains were well crystallized and the shell was amorphous. EDS analysis was used to check the composition of the core-shell with Ca, Cu, Ti and Al elements.

SEM investigation of the compounds after sintering has exhibited that the average size grain for coated samples (120 µm) is bigger than CCTO (70 µm) (**Figure 14a**). At the same time, the density is found better for CCT@$Al_2O_3$. Then, $Al_2O_3$ phase acts as a sintering aid. The appearance of a liquid phase, thanks to the CuO–$Al_2O_3$ interaction during the ceramic sintering heating stage, results in enhanced growth of ceramics grains and density. The presence of Al element and CuO phase in the GB has been attested by EDS analysis in CCTO@$Al_2O_3$ sample (**Figure 14b**). However, there is no significant observation of the Al element within the grains (area 1).

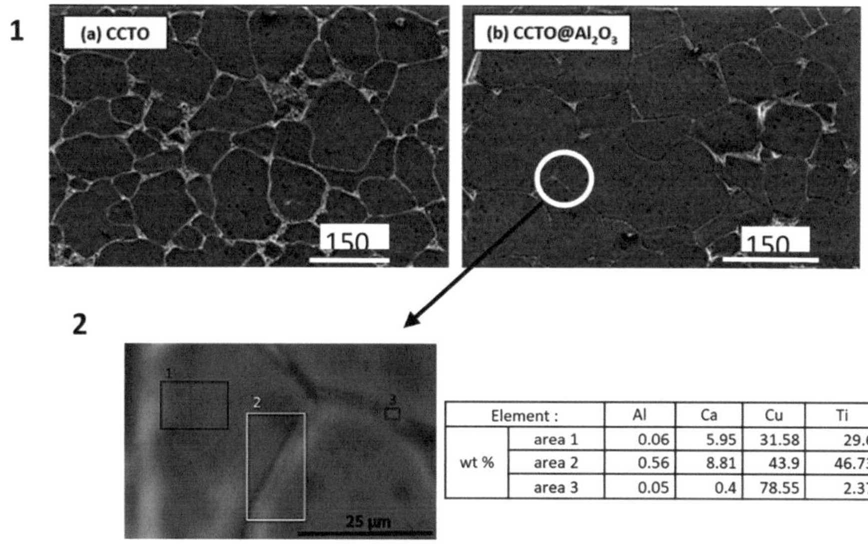


| Element : | | Al | Ca | Cu | Ti |
|---|---|---|---|---|---|
| wt % | area 1 | 0.06 | 5.95 | 31.58 | 29.6 |
| | area 2 | 0.56 | 8.81 | 43.9 | 46.73 |
| | area 3 | 0.05 | 0.4 | 78.55 | 2.37 |

**Figure 14.**
*(1) SEM images of (a) CCTO and (b) CCTO@$Al_2O_3$ pellets (2) EDS analysis in different areas: grain (1), grain boundary (2) and copper-rich phase (3) [61].*

Dielectric measurements on the synthesized nanocomposites indicated that the dielectric constant ($\varepsilon_r$) at 1 kHz is higher with a value close to 91,000 than CCTO (68000). This increase can be correlated with the IBLC model describing the mechanisms leading to colossal permittivity of CCTO [48]. An interesting point of these results is that the loss tangent is slightly lower but the resistivity is slightly increased. However, the functionalization of CCTO by PEG or PVP further improves the dielectric properties and, in particular, the bulk resistivity which increases by an order to achieve 2 $10^8$ Ω cm. Dielectric losses are linked to the modification of the interfacial polarization between the grains, which results from the accumulation and movement of charges at the interfaces. The presence of the compact insulating $Al_2O_3$ phase reduces the concentration and mobility of these carriers, leading to a reduction in dielectric losses and an increase in resistivity.

### 3.8 Titanium oxide-based core-shell nanocomposites

The coating of CCTO particles with a titanium shell and different thicknesses has also been applied with the same process used for $Al_2O_3$ with the appropriate precursor tetrabutoxyde de titane -TBOT (sigma Aldrich, 97%). The scanning and transmission electron microscopy clearly indicate the shell around CCTO grains. The SEM images show a shell around a grain agglomerate. TEM and STEM clearly indicate that the obtained samples are composed of well-crystallized CCTO and amorphous $TiO_2$. Dielectric measurements were realized on pellets sintered at 1050°C for 24 h, as previously done, to compare the impact of the shell's composition. The permittivity found for this series is very high, with a value of 870,000 at 1 kHz for CCT@$TiO_2$ (10 nm), at the expense of losses and bulk resistivity. The evolution and thermal stability of CCT@$TiO_2$ (10 nm) compound using in situ variable temperature chamber (HT-XRD) has been determined from ambient to 1000°C. Regarding the composition evolution of the phases, the crystallization of $TiO_2$ phase starts to form anatase structure for T = 600°C. As temperature rises, transformation to the rutile phase takes place at 850°C due to the isotropic thermal parameters change for oxygen. In addition, the rutile phase amount starts to rise continually. The existence of rutile phase in the CCTO@$TiO_2$ pellets was confirmed by Raman scattering in GBs. Baumard et al. have shown that under the experimental conditions of $TiO_2$ synthesis, it becomes conductive [76]. Thus it is possible that the rutile GB has become slightly conductive, which would explain the low GBs resistivity and the high tan(δ) found. Therefore, the overall resistivity of the sample that depends on the GB resistivity is also very low. If the GB is slightly conductive, the resistivity is reduced but if the GB is not sufficiently insulating, the electrical losses increase. Thus to improve the ceramic properties, one way would be to make insulator the GBs. It could be shown that the $TiO_2$ phase in anatase form is insulating and, therefore, improves the properties of grain joints [59]. However, the anatase form is metastable regardless of temperature and pressure, whereas rutile is an equilibrium polymorph, in agreement with CCTO@$TiO_2$.

## 4. Toward new compositions and core-shell design

The coating of CCTO grains with single composition shells ($SiO_2$, $TiO_2$, $Al_2O_3$) has been successfully achieved and the shell thickness obtained can be controlled. Permittivity decreased slightly for CCTO@$SiO_2$ and CCTO@$Al_2O_3$ but remains high ($\sim 10^4$), this means that the shell does not destroy the properties. Dielectric losses are improved for these both compounds. It has also been found that resistivities are improved and more specifically for $Al_2O_3$ coating with a bulk resistivity that increases from $10^6$ to $10^9$ Ω cm. While the dielectric constant is greatly improved by CCTO@$TiO_2$ linked to the GB capacitance, the bulk resistivity is low [77]. The coating of CCTO compared to additives or conventional doping leads to a decrease in dielectric losses while maintaining a high permittivity [78–80]. Taking into account the best properties of each compound for a shell thickness close to 10 nm, shell mix was considered, CCT@$SiO_2$@$TiO_2$, CCTO@$SiO_2$@$Al_2O_3$, CCT@$TiO_2$@$Al_2O_3$ and CCT@$SiO_2$@$Al_2O_3$@$TiO_2$. The objectives were (i) to keep the high dielectric constant by improving the bulk resistivity (ii) to observe the impact of several compositions on the dielectric properties (iii) to correlate the ε and the capacitance of GBs and the bulk resistivity and the GB resistivity to confirm the IBLC model. The process of synthesis of these powders remained the same as before by mixing the precursors TEOS, TBOT and Aluminum propoxide depending on the desired shell. The structural and microstructural characterizations (HRTEM-SEM) showed the good conformity of the shells around the grains and the amorphous character of the shell. The question arises as to whether the 3 shells form a solid solution when the ceramics are sintered. According to the phase diagrams of the 3 binary oxides, the mutual solubility of the oxides is negligible. In the ternary $Al_2O_3$-$TiO_2$-$SiO_2$ system, only the aluminosilicate compound $Al_6Si_2O_{13}$ could be formed at identical ratios [81]. However, for $T < 1600$°K, phase formation is negligible. The sintering temperature of 1050°C we used is too low, and no such phase has been demonstrated, only $TiO_2$ which crystallizes. From impedance spectroscopy, by fitting the data with two RC elements in series consisting of a resistor and capacitor connected in parallel, one for the grains and the second for GBs, the capacitance and resistivity of the grains and GBs were analyzed for all samples. As indicated by the IBLC model, the permittivity plotted as a function of the capacitance of the GBs follows a linear dependence in agreement with the Eq. (1) (**Figure 15a**). Moreover, GB resistivity is significantly more insulating than grain resistivity and as GB resistivity decreases, so does the overall sample resistivity is. In these figures, it

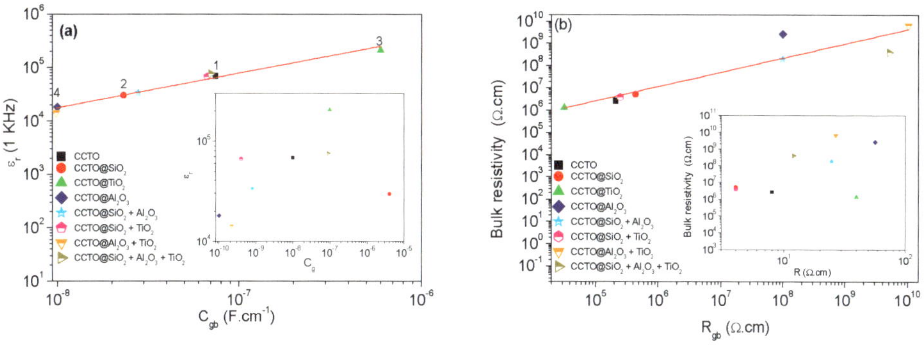


**Figure 15.**
*(a) Correlation of $\varepsilon_r$ and $C_{gb}$ for undoped CCTO (black dot) and "core-shell" CCTO samples. No correlation of $\varepsilon$ and $C_b$ for the same samples is found in the inset (b) Correlation of sample resistivity with $R_{gb}$ and its absence with $R_b$ shown in the inset.*

can be seen that a very high permittivity also shows the lowest resistivity, which becomes incompatible with the characteristics expected for a ceramic capacitor.

## 5. Toward an innovative cold plasma technology to coat dielectric particles

Cold plasmas, also named non-thermal plasmas, have shown their high efficiency in many industrial applications, including abatement of gaseous and liquid hazardous species [82–85], inactivation of micro-organisms in biomedical applications, food industry [86–91], and surface interactions [92–95]. In this type of plasma, chemical processes are initiated by collisions between energetic electrons (hot electrons) accelerated by the electric field and heavy species (atoms, molecules). Consequently, the gas temperature remains close to ambient, while the electronic temperature is much higher (some 10,000 K).

During the last 20 years, new promising processes based on various micro-plasma configurations were investigated [96–99]. Micro-plasmas are characterized by a reduced size. One of the discharge dimensions must be sub-millimeter, corresponding to a value of p × d (pressure × discharge gap) between 1 and 13 Pa m close to those of low-pressure plasmas, but with pressures that can reach atmospheric pressure. Consequently, current and energy densities are much higher than for conventional low-pressure plasmas. This leads to greater efficiency in the production of chemically active species in such plasmas [98]. These processes with small reaction volumes translate directly into short processing times.

Plasma Enhanced Chemical Vapor Deposition (PECVD) technique was successfully used for many years for plasma deposition processes. Classically, the plasma was generated by an RF-excitation (at 13.56 MHz) [100–102]. However, the coupling of a micro-plasma reactor with a nanosecond pulsed electric generator can present many advantages: on the one hand, this allows to benefit from the intrinsic micro-plasmas properties and on the other hand, the nanosecond-pulsed plasmas results in a very efficient energy transfer, i.e. input electric energy is mainly used to produce hot electrons leading to chemical reactions but not to gas heating.

This is why it is important to estimate excitation temperatures, *i.e.* rotational, vibrational and electronic temperatures (Trot, Tvib, Te) in order to characterize the plasma reactivity. These parameters provide information especially on the input energy distribution, and, contribute to identify the main mechanisms responsible for chemical transformations into the plasma.

For all these reasons, it seems interesting to explore a new approach to coat CCTO grains with silicon oxides by cold plasma treatments.

## 5.1 Experimental procedures

CCTO powder was prepared by standard solid-state reaction starting from a stoichiometric mixture of $CaCO_3$, CuO and $TiO_2$, described in detail above. Next, CCTO powder was mixed with an Ink Vehicle (Fuel Cells Materials) and deposited by screen-printing on a glass substrate, and finally ink.

was evaporated by heat treatment.

$SiO_2$ is obtained from the decomposition of hexamthyldisiloxane (HMDSO, $C_6H_{18}OSi_2$) vapor by an argon/oxygen cold plasma. A low HMDSO proportion in the Ar/$O_2$/HMDSO mixture is required to obtain a higher quality $SiO_2$ film, typically a few 100 ppm of HMDSO [103]. In addition, in order to have a complete oxidation of HMDSO to $SiO_2$, $CO_2$ and water vapor, a stoichiometric molecular ratio of $O_2$/HMDSO must be kept at a value of 13 [104]. Plasma treatments were performed into a 21 L vacuum chamber, the Ar/$O_2$/HMDSO mixture was introduced in the micro-torch by mains of controlled gas and liquid mass flow meters at a working pressure of 2.9 $10^3$ Pa. A compromise was found between thickness, layer homogeneity and processing time for the gas mixture composition and the time treatment: Ar/$O_2$/HMDSO = 2028 Ncm$^3$ mn$^{-1}$/7.84 Ncm$^3$ mn$^{-1}$/523 mg h$^{-1}$, respectively, and 30 min treatment in this study, the shell thickness being mainly controlled by the processing time.

The experimental setup used to realize the plasma treatments is illustrated in **Figure 16**. Optical emission measurements were performed with a 750 mm imaging

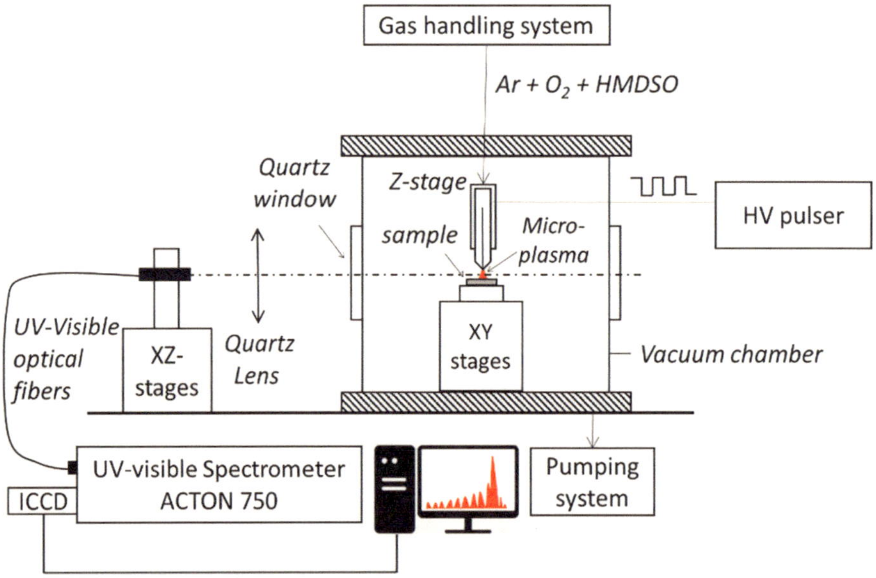


**Figure 16.**
*Experimental setup of the plasma generator system and spectroscopic acquisition.*

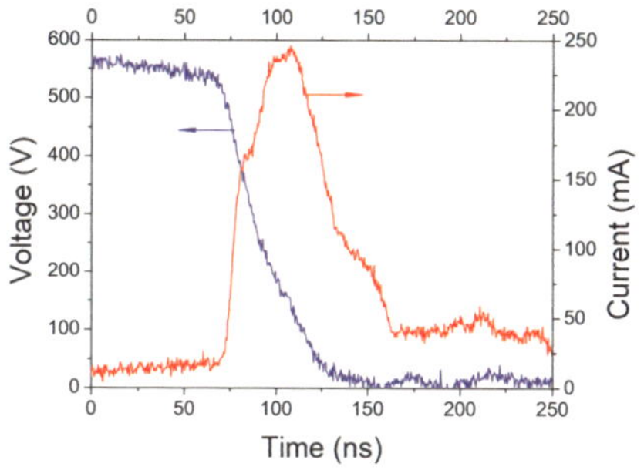


**Figure 17.**
*Time evolution of voltage and current of a single pulsed discharge (630 V applied voltage).*

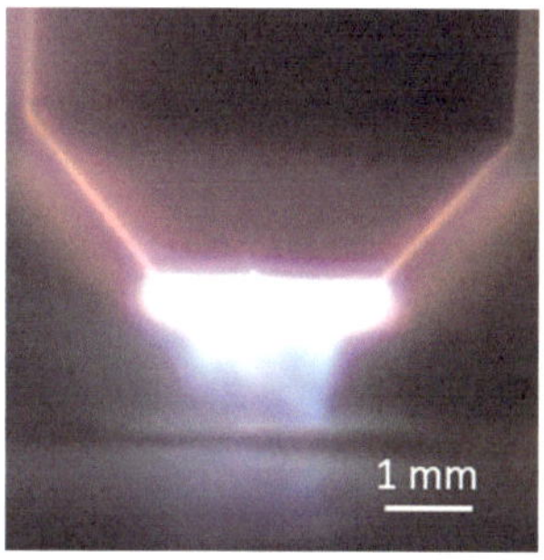


**Figure 18.**
*Profile of the micro-torch plasma.*

spectrometer equipped with an ICCD camera. The micro-torch was driven by a high voltage pulsed generator controlled by an arbitrary function generator set in pulse mode. The maximum of current was achieved in a very short time of about 30 ns, as it can be seen on example of voltage and current record of a single discharge displayed in **Figure 17**.

The micro-torch was a wire/cylinder configuration with an annular discharge gap of 250 μm and an outer diameter of 2 mm. The gas mixture flowed on the axis of the reactor and the plasma appeared as a needle going out of the reactor. The CCTO support was placed on a translational motorized XY stage, which allows to treat a surface of 15 × 15 $mm^2$, an example of the micro-torch plasma behavior is given in **Figure 18**.

## 6. Results and discussion

### 6.1 Spectroscopic investigations

Optical emission spectroscopy (OES) is a powerful, non-intrusive plasma diagnostic tool. It allows the identification of the species present in the plasma phase. In addition, a comparison between experimental and simulated molecular spectra provides access to the different plasma excitation temperatures. Emission of plasma was investigated in the UV–visible range. The main atomic species observed were Si I, C I, Ar I, Ar II, O I. The identified molecular species were the ro-vibrational bands of SiO

($A^1\Pi$-$X^1\Sigma^+$) at 234–250 nm, $OH^°$ ($A^2\Sigma^+$-$X^2\Pi_i$) (0,0) at 309 nm, CH ($B^2\Sigma$-$X^2\Pi$) (0,0) at 390 nm and CH ($A^2\Delta$-$X^2\Pi$) (0,0) at 430 nm, and the $C_2$ Swan band ($d^3\Pi_g$-$a^3\Pi_u$) (0,0) at 516.5 nm. An example is shown in **Figure 19**. The presence of atomic and diatomic species derived from the HMDSO molecules demonstrates the strong dissociation power of this type of plasma.

Several molecular spectral simulation software are available on the market. In this study, we have used *SPECAIR* and *LIFBASE* codes [105, 106]. Among other functions, they can be used to easily estimate plasma excitation temperatures by adjusting plasma parameters in order to obtain the best fit between simulated and experimental spectra.

An example of such processing is shown in **Figure 20** for estimating rotational temperature. The three ro-vibrational bands $OH^°$ ($A^2\Sigma^+$-$X^2\Pi_i$) (0,0) at 309 nm, CH ($B^2\Sigma$-$X^2\Pi$) (0,0) at 390 nm and CH ($A^2\Delta$-$X^2\Pi$) (0,0) at 430 nm were used as

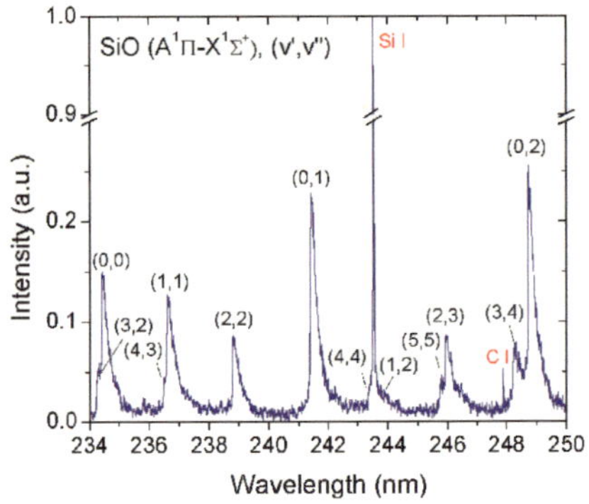


**Figure 19.**
*Experimental spectrum in the range of 234 nm to 250 nm, mainly composed of SiO emissions, with the presence of C I and Si I lines (1000 V applied voltage).*

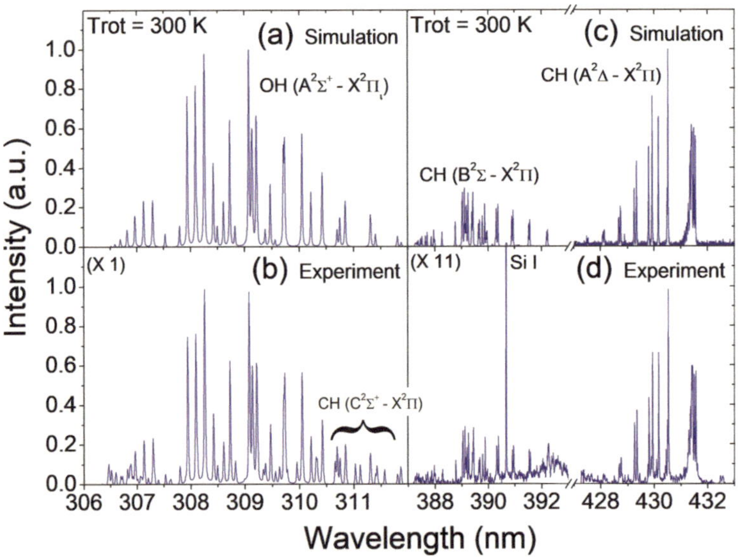


**Figure 20.**
*Spectra of $OH^°$ ($A^2\Sigma^+$-$X^2\Pi_i$), CH ($B^2\Sigma$-$X^2\Pi$) and CH ($A^2\Delta$-$X^2\Pi$). (a) and (c) LIFBASE simulated spectra, and (b) and (d) experimental spectra (1000 V applied voltage).*

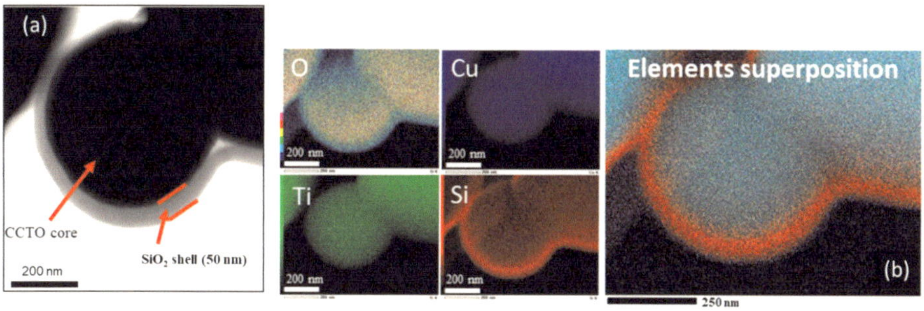


**Figure 21.**
*(a) TEM picture of a core-shell obtained with Ar/$O_2$/HMDSO = 2028 N $cm^3$ $mn^{-1}$/7.84 N$cm^3$ $mn^{-1}$/523 mg $h^{-1}$, respectively, 30 min treatment and 1000 V applied voltage and (b) EDS-STEM mapping pictures of elements and their superposition O, Cu, Ti, Si.*

thermometer species. The rotational temperature Trot, usually assimilated to the gas temperature Tg [107], was estimated individually for the three thermometers by simulated spectra adjustments, built from the *LIFBASE* code. The best adjustment was Trot = 300 ± 10 K.

The same technique can be used to estimate the vibrational temperature of molecules Tvib, as well as electron temperature Te. These values have been estimated under our experimental conditions at Tvib = 2400 K to 3700 K (± 100 K) (depending on the applied voltage) and Te = 5.3 eV ± 0.2 eV.

These results show that this micro-plasma produces highly non-Local Thermodynamic Equilibrium (non-LTE) or cold plasmas, and respects the characteristic inequality of cold plasmas:

$$T_{rot} = T_g < T_{vib} \ll T_e \quad (4)$$

### 6.2 Microstructure analysis

TEM analysis revealed the presence of a particularly homogeneous shell around the CCTO grains with a constant thickness, as shown in **Figure 21a**, where the thickness is 50 nm ± 2 nm for this sample. In addition, any orientation of $SiO_2$ shell was observed, which confers an amorphous character to the shell. X-ray analysis confirmed the shell's amorphous structure.

The mapping data of principal elements present in the shell zone have been analyzed by EDS-STEM and shown in **Figure 21b**. The mapping of O, Cu, Ti, Si and their superposition confirmed that CCTO grains are coated with Si. This means that there is an association between Si and O to obtain SiO ($SiO_2$) which confirms the core-shell structure.

## 7. Conclusion

Nanocomposites depicted in this review are composed of a calcium copper titanate (CCTO) core embedded in $SiO_2$, $Al_2O3$ or TiO2 shells. This strategy was used in order to meet the specifications in the industry, i.e. materials with a strong dielectric constant, low losses and a strong bulk resistivity. Two methods to achieve core-shell nanostructured materials have been described i) the chemical pathway and ii) the cold plasma pathway. For the chemical way, we focused on coating with different shell compositions by modifying the surface of CCTO grains, by modulating the shell thickness at the nanoscale. Thus the modification of ceramics could make an important contribution to the control of properties through the engineering of interfaces. The main objective of this study was to control grain boundaries which are essential for obtaining good dielectric and electrical properties. While "core-shell" materials have already proved their interest in controlling properties, there are still points to be improved to understand the phenomena. In particular, in-depth studies of grain boundaries structure at the atomic scale should be carried out to better understand chemical segregation, interdiffusion but also the investigation of the semi-conductor character of grains. The reduction of the sintering temperature would be a plus to limit the deterioration of shells and preserve this microstructure.

The cold plasma coating described in this work is advantageous over other methods reported in the recent literature like: sol-gel, Stöber, hétéro-coagulation [105–107] due to its homogeneous shell and controlled thickness, its process rapidness (max 1 h vs. 1–3 days resp.) and its dry character (no by-products). Another important advantage is being able to work on all surfaces and particles with different chemical composition of the deposition, compared to the layer-by-layer deposition method [74, 108]. The metalorganic precursor (HMDSO) is totally dissociated in the plasma, leading to the nucleation of $SiO_2$ on the surface of CCTO particles. This system facilitated the generation of uniform core-shell grains with nanometric-controlled shell thickness. The results of temperature measurements described in this study i.e. Trot < Tvib << Te, confirm the non-thermic character of this micro-plasma. This type of plasma is therefore well adapted to chemical treatments, by efficiently generating chemically active species by electronic collisions, without overheating the whole gas.

## Acknowledgements

Authors would like to thank J. Wolfman from GREMAN laboratory (France) for the TEM lamella preparation by FIB (Focus Ion Beam) and S. Magrex for final proofreading.

This work was funded in part by Région Centre-Val de Loire within the projects, Concerto (2012) and Coconut (2016).

## Author details

Cécile Autret-Lambert*, Samir Merad, Sonia De Almeida-Didry, Olivier Motret and François Gervais
GREMAN, UMR7347 CNRS, Université de Tours, Tours, France

*Address all correspondence to: cecile.autret@univ-tours.fr

## References

[1] Sakabe Y, Minai K, Wakino K. High-dielectric constant ceramics for base metal monolithic capacitors. Japanese Journal of Applied Physics. 1981;**20**:147

[2] Hao H, Liu M, Liu H, Zhang S, Shu X, Wang T, et al. Design, fabrication and dielectric properties in core-double shell $BaTiO_3$-based ceramics for MLCC application. RSC Advances. 2015;**5**: 8868-8876

[3] Zhang Y, Wang X, Kim J-Y, Tian Z, Fang J, Hur K-H, et al. High performance $BaTiO_3$-based BME-MLCC nanopowder prepared by aqueous chemical coating method. Journal of the American Ceramic Society. 2012;**95**:1628-1633

[4] Kishi H, Mizuno Y, Chazono H. Base-metal electrode-multilayer ceramic capacitors: Past, present and future perspectives. Japan Journal of Applied Physics. 2003;**42**:1-15

[5] Pithan C, Hennings D, Waser R. Progress in the synthesis of nanocrystalline $BaTiO_3$ powders for MLCC. International Journal of Applied Ceramic Technology. 2005;**2**:1-14

[6] Pan M-J, Randal CA. A brief introduction to ceramic capacitors. IEEE Insulation Electrical Management. 2010; **26**:44-50

[7] Hong J-Y, Lu H-Y. Polar nanoregions and dielectric properties of $BaTiO_3$-based Y5V multilayer ceramic capacitors. Materials Science. 2014;**97**:2256

[8] Pecharroman C, Esteban-Betegón F, BartolomØ JØF, López-Esteban S, Moya JØS. New percolative $BaTiO_3$-Ni composites with a high and frequency-independent dielectric constant ($\epsilon r \approx$ 80000). Advanced Materials. 2001;**13**:1541

[9] Waku S. High dielectric constant ceramic material and method of producting same. US Patent 3/473, 958. 1969

[10] Frechette MF, Trudeau ML, Alamdari HD, Boily S. Introductory remarks on nanodielectrics. IEEE Transactions on Dielectrics and Electrical Insulation. 2004;**11**:808-818

[11] Baig N, Kammakakam I, Falath W. Nanomaterials: A review of synthesis methods, properties, recent progress, and challenges. Materials Advance. 2021; **2**:1821-1871

[12] Park Y, Kim Y. The dielectric temperature characteristic of additives modified barium titanate having core-shell structured ceramics. Journal of Materials Research. 1995;**10**:2770

[13] Jeon Sang-Chae J. TCC behavior of a shell phase in core/shell structure formed in Y-doped $BaTiO_3$: An individual observation. Korean Crystal Growth and Crystalline Technology. 2020;**30**:110-116

[14] Elissalde C, Chung UC, Roulland F, Berthelot R, Artemenko A, Majimel J, et al. Specific core-shell approaches and related properties in nanostructured ferroelectric ceramics. Ferroelectrics. 2018;**532**:138-159

[15] Bi K, Bi M, Hao Y, Luo W, Cai Z, Wang X, et al. Ultrafine core-shell $BaTiO_3@SiO_2$ structures for nanocomposite capacitors with high energy density. Nano Energy. 2018;**51**: 513-523

[16] De Rase S, Roy R. Phase equilibria in the system $BaTiO_3$-$SiO_2$. Journal of American Ceramic Society. 1995;**38**: 389-395

[17] Robertson J. High dielectric constant gate oxides for metal oxide Si transistors. Report Progress in Physics. 2006;**69**: 327-396

[18] Hennings D, Janssen R, Reynen JL. Control of liquid-phase-enhanced discontinuous grain growth in barium titanate. Journal of American Ceramic Society. 1987;**70**:23-27

[19] Nakano Y. Microstructure and related phenomena of multilayer ceramic capacitors with Ni-electrode. Technical Report, DTIC Document. 1993

[20] Wei F, Burn I. Core-shell structured dielectric particles for use in multilayer ceramic capacitors. US Patent App. 12/291,153. 2008

[21] Prakash SB, Varma KBR. Influence of sintering conditions and doping on the dielectric relaxation originating from the surface layer effects in $CaCu_3Ti_4O_{12}$ ceramics. Journal of Physical and Chemical Solids. 2007;**68**:490-502

[22] Cao G, Feng L, Wang C. Grain-boundary and subgrain-boundary effects on the dielectric properties of $CaCu_3Ti_4O_{12}$ ceramics. Journal of Physics D: Applied Physics. 2007;**40**: 2899

[23] Wu Y, Zhu S, Park S, Shapiro S, Shirane G. Defect structure of the high-dielectric-constant perovskite $CaCu_3Ti_4O_{12}$. Journal of Tafto Physics Review B. 2005;**71**:014118

[24] Morrison FD, Coats AM, Sinclair DC, West AR. Charge compensation mechanisms in La-doped $BaTiO_3$. Journal of Electroceramics. 2001;**6**:219

[25] Autret-Lambert C, Merad S, De Almeida-Didry S, Roger S, Gervais F. Effect of MnO on the dielectric properties of Nb-$SrTiO_3$ ceramics for capacitor applications. ACS Applied Engineering Materials. 2023;**1**:68-79

[26] Chung C, Elissalde S, Mornet MM, Estournès C. Controlling internal barrier in low loss $BaTiO_3$ supercapacitors. Applied Physics Letters. 2009;**94**:072903

[27] Ramirez AP, Subramanian MA, Gardel M, Blumberg G, Li D, Vogt T, et al. Giant dielectric constant response in a copper-titanate. Solid State Communications. 2000;**115**:217-274

[28] Chung SY, Kim ID, Kang SJL. Strong nonlinear current-voltage behaviour in perovskite-derivative calcium copper titanate. Nature Materials. 2004;**3**:774-778

[29] Xu D, He K, Yu R, Jiao L, Yuan H, Sun X, et al. Effect of $AETiO_3$ (AE = Mg, Ca, Sr) doping on dielectric and varistor characteristics of $CaCu_3Ti_4O_{12}$ ceramic prepared by the sol-gel process. Journal of Alloys and Compounds. 2014;**592**: 220-225

[30] Subramanian MA, Li D, Duran N, Reisner BA, Sleight AW. High dielectric constant in $ACu_3Ti_4O_{12}$ and $ACu_3Ti_3FeO_{12}$ phases. Journal of Solid State Chemistry. 2000;**151**:323-325

[31] Mauczok R, Wernicke R. Ceramic boundary-layer capacitors. Philips Technology Review. 1983;**41**:338-346

[32] Chiou BS, Lin ST, Duh JG, Chang PH. Equivalent circuit model in grain-boundary barrier layer capacitance. Journal of American Ceramic Society. 1989;**72**:1967-1975

[33] Marchin L, Guillemet-Fritsch S, Durand B, Levchenko AA, Navrotsky A, Lebey T. Grain growth-controlled giant permittivity in soft chemistry $CaCu_3Ti_4O_{12}$ ceramics. Journal of the American Ceramic Society. 2008;**91**: 485-489

[34] Bochu B, Deschizeaux MN, Joubert JC, Collomb A, Chenavas J, Marezio M. Synthèse et caractèrisation d'une sèrie de titanates pèrowskites isotypes de $[CaCu_3](Mn_4)O_{12}$. Journal of Solid State Chemistry. 1979;**29**:291-298

[35] Moussa SM, Kennedy BJ. Structural studies of the distorted perovskite $Ca_{0.25}Cu_{0.75}TiO_3$. Materials Research Bulletin. 2001;**36**:2525-2529

[36] Bodeux R, Gervais M, Wolfman J, Autret-Lambert C, Liu G, Gervais F. $CaCu_3Ti_4O_{12}$ thin film capacitors: Evidence of the presence of a Schottky type barrier at the bottom electrode. Thin Solid Films. 2012;**520**:2632-2638

[37] Kant C, Rudolf T, Mayr F, Krohns S, Lunkenheimer P, Ebbinghaus SG, et al. Broadband dielectric response of $CaCu_3Ti_4O_{12}$: From dc to the electronic transition regime. Physical Review B. 2008;**77**:045131

[38] Krohns S, Lu J, Lunkenheimer P, Brizé V, Autret-Lambert C, Gervais M, et al. Correlations of structural, magnetic, and dielectric properties of undoped and doped $CaCu_3Ti_4O_{12}$. The European Physical Journal B. 2009;**72**: 173-182

[39] Homes CC, Vogt T, Shapiro SM, Wakimoto S, Subramanian MA, Ramirez AP. Charge transfer in the high dielectric constant materials $CaCu_3Ti_4O_{12}$ and $CdCu_3Ti_4O$. Physical Review B. 2003;**67**:092106-092104

[40] Sinclair DC, Adams TB, Morrison FD, West R. $CaCu_3Ti_4O_{12}$: One-step internal barrier layer Capacitor. Applied Physics Letters. 2002; **80**:2153-2155

[41] Yang J, Shen M, Fang L. The electrode/sample contact effects on the dielectric properties of the $CaCu_3Ti_4O_{12}$ ceramic. Materials Letters. 2005;**59**: 3990-3993

[42] Yu R, Xue H, Cao Z, Chen L, Xiong Z. Effect of oxygen sintering atmosphere on the electrical behavior of CCTO ceramics. Journal of the European Ceramic Society. 2012;**32**:1245-1249

[43] Yang Y, Wang X, Liu B. $CaCu_3Ti_4O_{12}$ ceramics from different methods: microstructure and dielectric. Journal of Material Science. 2014;**25**:146-151

[44] Adams TB, Sinclair DC, West AR. Influence of processing conditions on the electrical properties of $CaCu_3Ti_4O_{12}$ ceramics. Physical Review B: Condensed Matter. 2006;**73**:094124

[45] Brize V, Gruener G, Wolfman J, Fatyeyeva K, Tabellout M, Gervais M, et al. Grain size effects on the dielectric constant of $CaCu_3Ti_4O_{12}$ ceramics. Material Science Engineering B. 2006; **129**:135

[46] Douy A. Polyacrylamide gel: An efficient tool for easy synthesis of multicomponent oxide precursors of ceramics and glasses. International Journal of Inorganic Materials. 2001;**3**:699

[47] Bender BA, Pan MJ. The effect of processing on the giant dielectric properties of $CaCu_3Ti_4O_{12}$. Material Science and Engineering B. 2005;**117**: 339-347

[48] De Almeida-Didry S, Autret C, Lucas A, Honstettre C, Pacreau F, Gervais F. Leading role of grain boundaries in colossal permittivity of doped and undoped CCTO. Journal of the European Ceramic Society. 2014;**34**: 3649-3654

[49] Rhouma S, Megriche A, El Amrani M, Said S, Roger S,

Autret-Lambert C. Effect of Sr/Mg co-doping on the structural, dielectric, and electrical properties of $CaCu_3Ti_4O_{12}$ ceramics. Journal of Materials Science: Materials in Electronics. 2022;**33**: 4535-4549

[50] Ahmadipour M, Ain MF, Ahmad ZA. A short review on copper calcium titanate (CCTO) electroceramic: Synthesis, dielectric properties, film deposition, and sensing application. Nano-Micro Letters. 2016;**8**:291-311

[51] De Almeida S, Autret C, Gervais F. Promising Trends in Materials for Ceramic Capacitors. London, United Kingdom: LAP Lambert Academic Publishing, Inc; 2020

[52] Li M, Feiteira A, Sinclair D. Influence of Mn doping on the semiconducting properties of $CaCu_3Ti_4O_{12}$ ceramics. West Applied Physics Letters. 2006;**88**:232903

[53] Kim CH, Jang YH, Seo SJ, Song CH, Son JY, Yang YS, et al. Effect of Mn doping on the temperature-dependent anomalous giant dielectric behavior of $CaCu_3Ti_4O_{12}$. Physical Review B. 2012; **85**:245210

[54] Yang Z, Zhang Y, You G, Zhang K, Xiong R, Shi J. Dielectric and electrical transport properties of the $Fe^{3+}$-doped $CaCu_3Ti_4O_{12}$. Journal of Material Science and Technology. 2012;**28**: 1145-1150

[55] Shao SF, Zhang JL, Zheng P, Wang CL. Effect of Cu-stoichiometry on the dielectric and electric properties in $CaCu_3Ti_4O_{12}$ ceramics. Solid State Communication. 2007;**142**:281-286

[56] Kim K, Lee J, Lee K, Kim D, Riu D, Lee SB. Microstructural evolution and dielectric properties of Cu-deficient and Cu-excess $CaCu_3Ti_4O_{12}$ ceramics. Material Research Bulletin. 2008;**43**: 284-291

[57] Liu L, Fan H, Chen X, Fang P. Electrical properties and microstructural characteristics of nonstoichiometric $CaCu_{3x}Ti_4O_{12}$ ceramics. Journal of Alloys and Compounds. 2009;**469**:529-534

[58] Fang TT, Hsu-Kai S. Mechanism for developing the boundary barrier layers of $CaCu_3Ti_4O_{12}$. Journal of the American Ceramic Society. 2004;**87**:2072-2079

[59] Fang TT, Mei L-T, Ho HF. Effects of Cu stoichiometry on the microstructures, barrier-layer structures, electrical conduction, dielectric responses, and stability of $CaCu_3Ti_4O_{12}$. Acta Materialia. 2006;**54**: 2867-2875

[60] De Almeida-Didry S, Autret C, Honstettre C, Lucas A, Zaghrioui M, Pacreau F, et al. Central role of $TiO_2$ anatase grain boundaries on resistivity of $CaCu_3Ti_4O_{12}$-based materials probed by Raman spectroscopy. Solid State Science. 2016;**61**:102-105

[61] De Almeida-Didry S, Nomel MM, Autret C, Honstettre C, Lucas A, Pacreau F, et al. Control of grain boundary in alumina doped CCTO showing colossal permittivity by core-shell approach. Journal of European Ceramic Society. 2018;**38**: 3182-3187

[62] Domengès B, Riquet G, Marinel S, Harnois C. High-resolution FIB-TEM-STEM structural characterization of grain boundaries in the high dielectric constant perovskite $CaCu_3Ti_4O_{12}$. Journal of the European Ceramic Society. 2020;**40**:3577-3584

[63] Stöber W, Fink A, Bohn E. Controlled growth of monodisperse

silica spheres in the micron size range. Journal of Colloid and Interface Science. 1968;**26**:62-69

[64] Maurice V, Georgelin T, Siaugue J-M, Cabuil V. Synthesis and characterization of functionalized core-shell $\gamma Fe_2O_3$-$SiO_2$ nanoparticles. Journal of Magnetic Materials. 2009;**321**: 1408-1413

[65] Congcong X, Zhipeng RS, Yaodong W. Tuning the microstructure of $BaTiO_3$@$SiO_2$ core-shell nanoparticles for high energy storage composite ceramics. Journal of Alloys and Compounds. 2019;**784**:173-181

[66] Deng T-S, Marlow F. Synthesis of monodisperse polystyrene@vinyl-$SiO_2$ core-shell particles and hollow $SiO_2$ spheres. Chemistry of Materials. 2012; **24**:536-542

[67] Celebi H, Duran S, Dogan A. The effect of core-shell $BaTiO_3$@$SiO_2$ on the mechanical and dielectric properties of PVDF composites. Polymer-Plastics Technology and Materials. 2022;**61**: 1191-1203

[68] Zhu X, Wang J, Zhang Z, Zhu J, Zhou S, Liu Z, et al. Perovskite nanoparticles and nanowires: Microwave-hydrothermal synthesis and structural characterization by high-resolution transmission electron microscopy. Journal of the American Ceramic Society. 2008;**91**: 2683-2689

[69] Jana NR, Earhart C, Ying JY. Synthesis of water-soluble and functionalized nanoparticles by silica coating. Chemistry of Materials. 2007;**19**: 5074-5082

[70] Hai C, Inukai K, Takahashi Y, Izu N, Akamatsu T, Itoh T, et al. Surfactant-assisted synthesis of mono-dispersed cubic $BaTiO_3$ nanoparticles. Materials Research Bulletin. 2014;**57**:103-109

[71] Yuan W. Investigation on the decomposable process and the secondary liquid phase effect on the dielectric properties of $CaCu_3Ti_4O_{12}$ ceramics. Journal of Physics D: Applied Physics. 2009;**42**:175401

[72] Rahimabady M, Mirshekarloo MS, Yao K, Lu L. Dielectric behaviors and high energy storage density of nanocomposites with core-shell $BaTiO_3$@$TiO_2$ in poly (vinylidene fluoride-hexafluoropropylene). Physical Chemistry Chemical Physics. 2013;**15**: 16242-16248

[73] He D, Wang Y, Chen X, Deng Y. Core-shell structured $BaTiO_3$@ $Al_2O_3$ nanoparticles in polymer composites for dielectric loss suppression and breakdown strength enhancement. Composites: Part A. 2017;**93**:137-143

[74] Peng H-P, Liang R-P, Qiu J-D. Facile synthesis of $Fe_3O_4$@$Al_2O_3$ core-shell nanoparticles and their application to the highly specific capture of heme proteins for direct electrochemistry. Biosensors and Bioelectronics. 2011;**26**:3005-3011

[75] Zhang W, Chi ZX, Mao WX, Lv RW, Cao AM, Wan LJ. One-nanometer-precision control of $Al_2O_3$ nanoshells through a solution-based synthesis route. Chemistry International. 2014;**53**: 12776-12780

[76] Baumard JF, Gervais F. Plasmon and polar optical phonons in reduced rutile $TiO_{2-x}$. Physical Review B. 1977;**15**: 2316-2327

[77] De Almeida-Didry S, Merad S, Autret-Lambert C, Nomel MM, Lucas A, Gervais F. A core-shell synthesis of $CaCu_3Ti_4O_{12}$ (CCTO) ceramics showing colossal permittivity and low electric

losses for application in capacitors. Solid State Sciences. 2020;**109**:106431

[78] Prakash BS, Varma KBR. Effect of the addition of $B_2O_3$ and $BaO$-$B_2O_3$-$SiO_2$ glasses on the microstructure and dielectric properties of giant dielectric constant material $CaCu_3Ti_4O_{12}$. Journal of Solid State Chemistry. 2007;**180**:1918-1927

[79] Thongbai P, Jumpatam J, Yamwong T, Maensiri S. Effects of $Ta^{5+}$ doping on microstructure evolution, dielectric properties and electrical response in $CaCu_3Ti_4O_{12}$ ceramics. Journal of European Ceramic Society. 2012;**32**:2423-2430

[80] Ren L, Yang L, Xu C, Zhao X, Liao R. Improvement of breakdown field and dielectric properties of $CaCu_3Ti_4O_{12}$ ceramics by Bi and Al co-doping. Journal of Alloys and Compounds. 2018;**768**: 652-658

[81] Ilatovskaia M, Bärtel F, Fabrichnaya O. Phase relations in the $Al_2O_3$-$TiO_2$-$SiO_2$ system. Ceramics International. 2020;**46**:29402-29412

[82] Li S, Dang X, Yu X, Abbas G, Zhang Q, Cao L. The application of dielectric barrier discharge non-thermal plasma in VOCs abatement: A review. Journal of Chemical Engineering. 2020; **388**:124275

[83] Schiavon M, Torretta V, Casazza A, Ragazzi M. Non-thermal plasma as an innovative option for the abatement of volatile organic compounds: A review. Water, Air, and Soil Pollution. 2017;**228**: 388

[84] Van Durme J, Dewulf J, Leys C, Langenhove HV. Combining non-thermal plasma with heterogeneous catalysis in waste gas treatment: A review. Journal of Applied Catalysis B: Environmental. 2008;**78**:324

[85] Vandenbroucke AM, Morent R, De Geyter N, Leys C. Decomposition of trichloroethylene with plasma-catalysis: A review. Journal of Hazardous Materials. 2011;**195**:30

[86] Tabares FL, Junkar I. Cold plasma systems and their application in surface treatments for medicine. Molecules. 2021;**26**:1903

[87] Pankaj SK, Wan Z, Keener KM. Effects of cold plasma on food quality: A review. Food. 2018;**7**:4

[88] Laroussi M. Nonthermal decontamination of biological media by atmospheric-pressure plasmas: Review, analysis, and prospects. IEEE Transactions on Plasma Science. 2002; **30**:1409

[89] Laroussi M, Akan T. Arc-free atmospheric pressure cold plasma jets: A review. Plasma Process and Polymers. 2007;**4**:777

[90] Moreau M, Orange N, Feuilloley MGJ. Non-thermal plasma technologies: New tools for bio-decontamination. Biotechnology Advances. 2008;**26**:610

[91] Surowsky B, Schlüter O, Knorr D. Interactions of non-thermal atmospheric pressure plasma with solid and liquid food systems: A review. Food Engineering Reviews. 2015;**7**:82

[92] Dimitrakellis P, Gogolides E. Hydrophobic and superhydrophobic surfaces fabricated using atmospheric pressure cold plasma technology: A review. Advances in Colloid and Interface Science. 2018;**254**:1

[93] Kogelschatz U. Dielectric-barrier discharges: Their history, discharge physics, and industrial applications.

Plasma Chemistry and Plasma Processing. 2003;**23**:1

[94] McCurdy PR, Butoi CI, Williams KL, Fisher ER. Surface interactions of $NH_2$ radicals in $NH_3$ plasmas. Journal of Physical Chemistry B. 1999;**103**:6919

[95] Dorai R, Kuchner MJ. A model for plasma modification of polypropylene using atmospheric pressure discharges. Journal of Physics D: Applied Physics. 2003;**36**:666

[96] Papadakis AP, Rossides S, Metaxas AC. Microplasmas: A review. The Open Applied Physics Journal. 2011; **4**:45

[97] Belmonte T, Arnoult G, Henrion G, Gries T. Nanoscience with non-equilibrium plasmas at atmospheric pressure. Journal of Physics D: Applied Physics. 2011;**44**:1

[98] Foest R, Schmidt M, Becker K. Microplasmas, an emerging field of low-temperature plasma science and technology. International Journal of Mass Spectroscopy. 2006;**248**:87

[99] Becker KH, Schoenbach KH, Eden JG. Microplasmas and applications. Journal of Physics D: Applied Physics. 2006;**39**:55

[100] Martinu L, Poitras D. Plasma deposition of optical films and coatings: A review. Journal of Vacuum Science and Technology A. 2000;**18**:2619

[101] Meyyappan M, Delzeit L, Cassell A, Hash D. Carbon nanotube growth by PECVD: A review. Plasma Sources Science and Technology. 2003;**12**:205

[102] Aparna G, Sharma CS, Pandey AK. Synthesis of patterned vertically aligned carbon nanotubes by PECVD using different growth techniques: A review. Journal of Nanoscience and Nanotechnology. 2017;**17**:2256

[103] Bellel A, Sahli S, Ziari Z, Raynaud P, Segui Y, Escaich D. Wettability of polypropylene films coated with $SiO_x$ plasma deposited layers. Surface and Coatings Technology. 2006;**201**:129

[104] Barni R, Zanini S, Riccardi C. Characterization of the chemical kinetics in an $O_2$/HMDSO RF plasma for material processing. Advances in Physical Chemistry. 2012;**2012**:1

[105] Laux CO, Spence TG, Kruger CH, Zare RN. Optical diagnostics of atmospheric pressure air plasmas. Plasma Sources Science and Technology. 2003;**12**:125-138

[106] Luque J, Crosley DR. Lifbase: Database and Spectral Simulation Program (version 2.1.1). CA, USA: SRI International; 1999

[107] Motret O, Hibert C, Pellerin S, Pouvesle JM. Rotational temperature measurements in atmospheric pulsed dielectric barrier discharge-gas temperature and molecular fraction effects. Journal of Physics D: Applied Physics. 2000;**33**:1493

[108] Zhang X, Chen H, Zhang H. Layer-by-layer assembly: From conventional to unconventional methods. Journal of Chemical Communications. 2007;**14**: 1395

Chapter 6

# Research of Multifunctional Ceramic Materials for Their Application

*Angelina Stoyanova-Ivanova and Stanislav Slavov*


## Abstract

A new challenge is obtaining and researching ceramic multifunctional materials containing phases with various properties, as well as Aurivillius phases, which determine their application. They show potential for use in electrochemical applications and ferroelectric and piezoelectric devices, sensors, and non-volatile memories. Presented are our studies of volumetric nonmonophasic ceramics from the RE-Ba-Cu-O (ReBCO, RE = *rare-earth; Y, Dy*) and Bi-Pb-Sr-Ca-Cu-O (B(Pb) SCCO) systems that are superconductors, obtained via solid phase synthesis. A bulk ceramic composite $Y1_{23}/BaCuO_2$ was synthesized with starting stoichiometry of 1:3:4(Y:Ba:Cu) *via* a one-step procedure. It has superconducting and magnetic properties at low temperatures. DyBCO bulk ceramic with a nano-$Fe_3O_4$ additive was synthesized and characterized to identify the phase and elemental composition, the microstructure, and the superconducting transition temperature. The Aurivillius phases were synthesized via solid-phase synthesis and a melt-quench method. B(Pb)SCCO ceramics (2223, 2212, and 2201), with conductive properties, have been used as an addition to the active mass of a Zn electrode. The method of mixing the materials was also investigated. Their behavior in an alkaline environment and positive influence on the properties and longevity of the nickel-zinc battery has been studied. Part of the obtained ceramic systems was patented.


**Keywords:** multifunctional ceramics, REBCO, B(Pb)SCCO, Aurivillius phases, synthesis electrochemical applications, synthesis

## 1. Introduction

In recent years, particular attention has been given to the synthesis and characterization of multifunctional materials, such as ceramics with various structures obtained through different methods, aiming at their practical applications. In the present paper, we present our previous results in the research of ceramic multifunctional materials which contain phases with various properties (ferromagnetic magnetic phases, superconductive ceramic phases, and others), as well as Aurivillius phases (which possess high-temperature dielectric properties). Conductive ceramics



IntechOpen

are a class of functional materials with high electrical conductivity and chemical stability [1–4]. Ever since the discovery of high-temperature conductance in 1987 by Chu and Wu [5], there has been a special interest in cuprate ceramics, the methods of their preparation, and their application.

The properties of these ceramics make them suitable for applications in items such as magnetic bearings, permanent magnets, power cables, etc. "High-temperature superconductor" is a term that has been used interchangeably with "cuprate superconductor" for compounds like yttrium barium copper oxide (YBCO) and bismuth strontium calcium copper oxide (BSCCO), however, there are also other high-temperature superconductors. Research shows that Re123 (Re = rare earth element, such as Y, Eu, Gd, Dy, Nd, S, Ho, Er) materials are known as superconductors with a high Tc [6]. The crystal structure of the ReBCO materials is a multilayered perovskite structure [7]. Research over the years has been mainly focused on finding techniques that increase the transition temperature, Tc. Researchers have created many compounds from the YBCO family to obtain a Tc that is higher than that of Y123 (92 K), such as Y124 with a Tc = 80 K [8] and Y247 with a Tc that can vary from 30 K to 95 K, depending on the oxygen content [9, 10]. In 2009 Aliabadi, Farshchi and Akhavan [11] and Tavana [12] discovered a new Y-based high-temperature superconductor in the face of Y-Ba-Cu-O (Y-358), which becomes superconductive at over 100 K.

$YBa_2Cu_3O_7$-x (Y123) and $Bi_2Sr_2Ca_2Cu_3O_7$ (BSCCO) remain some of the most investigated superconductive systems. A reproducible and well-defined ceramic compound with nominal composition (Y123/$BaCuO_2$) has been synthesized in a one-step reaction. It has been found that barium cuprate is one of the few copper oxides exhibiting ferromagnetic interactions, thus during Tc measurements, it can affect the transition width of the ρ-T curves [13, 14]. Studies have shown that, besides the superconducting properties, a composite ReBCO (Re = rare earth) compound that contains $BaCuO_2$ exhibits magnetic properties as well. Thus, one can expect that such a diversity in the non-monophasic ReBCO ceramic's property might be useful for future practical application [15, 16].

Using solid phase synthesis a bulk sample Dy-123 doped with nano-$Fe_3O$ was obtained. Inserting a low concentration of $Fe_3O_4$ into a Dy-Ba-Cu-O sample improves its superconductive properties. We assume that this occurs due to the presence of $Fe_3O_4$ nanoparticles and $BaCuO_2$ and CuO phases at the boundaries of the grain which improves its connectivity [17].

The stability of these ceramics in a strong alkaline environment has been researched and its influence on their structure and properties has been traced. The magnetic properties of the resulting samples were measured, both before and after the electrochemical testing. The results show that the samples retain their superconductive properties after a stay in a strongly alkaline environment. They can be successfully used as an additive to the electrode in alkaline batteries, thus improving their cyclic operation capabilities [18].

Previous research conducted by us also demonstrates the potential of superconducting ceramics such as B(Pb)SCO 2201 and B(Pb)SCCO 2212 to be used as additives to the active mass of the zinc electrode in batteries and the effect on the electrochemical properties have been presented [19–27]. The influence of the different methods of obtaining a Zn active mass has been studied, as well as the content of the additive inside it [24, 25]. Using impedance spectroscopy the obtained Ni-Zn electrochemical alkaline systems (anode of nano-sized ZnO doped with differing content amounts of conducting ceramics) have been studied [26, 27]. The observed effect of adding them improves the battery's properties and lifespan, which makes B(Pb)SCCO cuprate ceramics a promising doping material for the development of new electrochemical systems, as well as for use in different devices.

The development of new materials with ferroelectric or hybrid properties, the following application directions can be defined as storage devices such as FERAM and DRAM and semiconductor elements [28], capacitors, sensors; high-temperature piezoelectric materials, namely the class of layered bismuth-containing compounds with additives, such as $SrBi_4Ti_4O_{15}$ with a Curie temperature of about 530°C [29] and $K_{0.5}Bi_{4.5}Ti_4O_{15}$ [30] with Curie temperature 555°C and d33 about 21.2 pC/N, Saito at al [31], Takenaka at al [32], Zhang et al. [33] with equivalent properties to piezoelectric bismuth-based materials, such as similar strain rate at room temperature, better thermal resistance up to the Curie temperature; electro-optical ceramic materials and their application — transparent thin films, usually the compounds $LiNbO_3$ and $Li(Nb, Ta)O_3$, and with an amorphous matrix with included ferroelectric crystals such as $(Na, K)NbO_3$, $BaTiO_3$ [34, 35], $LiNbO_3$ [36], $Bi_2VO_{5.5}$ [37], $Bi_2GeO$ [38]. The presented results regarding multifunctional ceramic materials show great potential for use in a wide spectrum of applications.

## 2. Materials and methods

### 2.1 Materials

The cuprate superconductors, also called cuprate ceramics, have been obtained *via* solid-phase synthesis. For the preparation of the composite material, a mixture of yttrium-barium-copper-oxide (YBCO) is obtained with an initial nominal stoichiometric composition of 1:2:3 and 1:3:4. Y2O3, BaCO3, and CuO are mixed and homogenized to form a homogeneous mixture. This mixture is then subjected to a three-step thermal treatment in an oxygen-rich environment [15, 16]. Following the same scheme, samples of the Dy-Ba-Cu-O system, which has been doped with nano-$Fe_3O_4$ [17, 39].

B(Pb)SCO 2201 and B(Pb)SCCO 2212 ceramics were prepared using a two-step solid-state synthesis. Starting materials including $Bi_2O_3$, PbO, CuO, $SrCO_3$, and $CaCO_3$ are dried, weighed in the required stoichiometric ratios, and homogenized. They are then heated at 780°C for 24 hours, crushed again, and pressed into tablets at 5–6 MPa. The tablets are subsequently fired: B(Pb)SCO 2201 at 830°C for 24 hours, and B(Pb)SCCO 2212 at 830°C for 48 hours under ambient air atmosphere [19].

To prepare the Zn electrode, an active mass is used, which is prepared from nanosized ZnO, carboxymethylcellulose (CMC 3% solution), polytetrafluoroethylene (PTFE 60% suspension) in a weight ratio of 81:14:5, and an additive B(Pb)SCO 2201 or B(Pb)SCCO 2212) with a different content [19–27]. CMC and PTFE serve as plasticizers. The mixing is done either through mechanical mixing, ultrasonic mixing, or ball milling. The prepared paste is applied onto a current collector made of modified copper foam. The obtained electrodes are dried at 70°C for one hour and pressed at a pressure of 30 MPa. The studied electrode package consists of electrodes measuring 5.0 × 3.0 cm with a thickness of 0.15 cm, separated by a microporous separator, and immersed in an alkaline electrolyte.

Polyphase samples were synthesized in the system $Bi_2O_3$–$TiO_2$–$Nd_2O_3$ using the method of melt quenching, and developed a methodology for the control of the constituent material Auriuvilius and pyrochlore crystalline phases. The method is used to control the high-frequency dielectric characteristics [40]. Additionally, the glass-crystal materials for sensors in the system $SiO_2$-$Bi_2O_3$-$TiO_2$ have been synthesized by a method of melt quenching and controlled crystallization of the glass and subsequent synthesis of thin layers by ink-screen-printing technique [41]. The materials in the system $La_2O_3$-$Gd_2O_3$-PbO-MnO-$B_2O_3$, demonstrate the synthesis of a monophasic material with a content of perovskite crystalline phase $(La_{1-x}Gd_x)_{0.6}$ $Pb_{0.4}MnO_3$ which has been shown to be promising for ferroelectric applications [42].

### 2.2 Methods

The following contemporary methods were used to analyze the structure and morphology of the synthesized and alkaline-treated samples: standard X-ray diffraction (XRD), scanning electron microscopy (SEM), and energy-dispersive X-ray spectroscopy (EDX). The non-stoichiometric oxygen coefficient ($\delta$) and the total oxygen content (y) were determined by a spectrophotometric method [43]. To establish the superconductive properties, magnetic measurements (AC/DC) were used.

For the electrochemical characterization of the initial ceramics and the obtained electrodes, the following techniques were employed: cyclic voltammetry (CV), chrono-potentiometry (CP), and potentiostatic electrochemical impedance spectroscopy (PEIS) using a Biologic potentiostat/galvanostat SP200 instrument.

## 3. Results and discussion

A bulk ceramic composite Y123/$BaCuO_2$ was synthesized by a solid-state reaction with starting stoichiometry of 1:3:4(Y:Ba:Cu) [15]. The ceramic composite Y123/ $BaCuO_2$ obtained in a one-step procedure was studied and appears to lead to a material possessing both superconducting and magnetic properties at low temperatures. We have shown that if the nominal starting composition is 1:3:4 (Y:Ba:Cu) one can obtain a two-phase composite with 62 wt% of superconducting YBCO with high critical temperature equal to 92.6 K and 38 wt% of magnetic $BaCuO_2$. It was also proven that the stability of obtained composite in alkaline solution is superior over pure Y123 material which is important for practical application.

The composite material was obtained which consists of a Y123 phase with oxygen content of 6.88 to 6.99 and a $BaCuO_2$ phase in a 2:1 ratio [15, 16]. This composite material and the method of obtaining it have been patented [16]. On **Figure 1** the x-ray diffraction spectra are presented: YBCO material with a superconductive Y123 phase (specter a), the obtained composite material containing a superconductive Y123 phase and a $BaCuO_2$ phase before (specter b) and after alkaline treatment (specter c), while **Figure 2** represents its magnetic hysteresis before (curve a) and after alkaline treatment (curve b). It is superconductive at a temperature higher than that of liquid nitrogen, possesses magnetic properties, and is resistant to alkaline environments. The presence of superconductive and magnetic properties and its resistance to alkaline environments increase the possibilities of application in all areas of technology.

Another composite material with multifunctional properties from the Dy-Ba-Cu-O (with a nominal composition of 123) system was also obtained by solid-phase

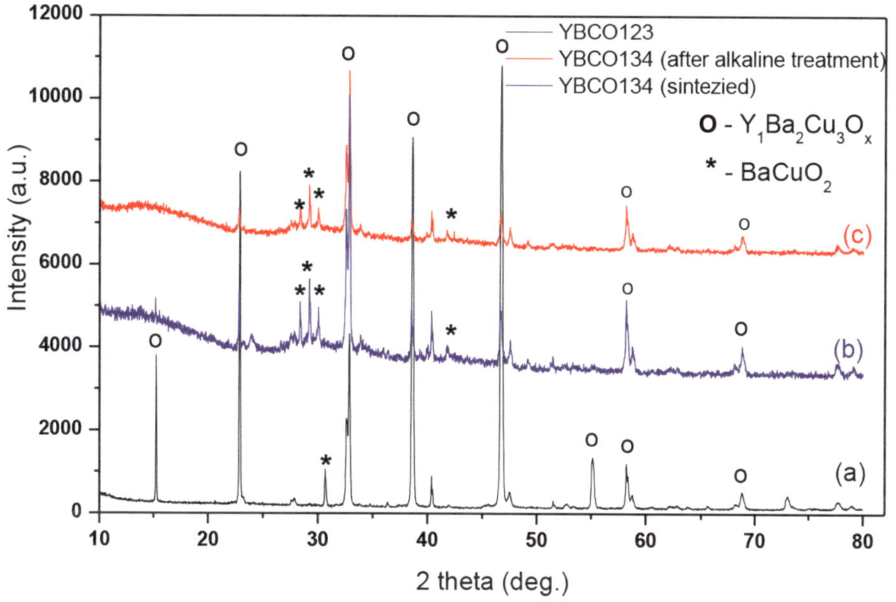


**Figure 1.**
*X-ray diffraction spectra of, respectively: YBCO material with a superconductive Y123 phase (spectrum (a)), the obtained composite material containing superconductive Y123 phase and $BaCuO_2$ phase before (spectrum (b)) and after its alkaline treatment (spectrum (c)) [16].*

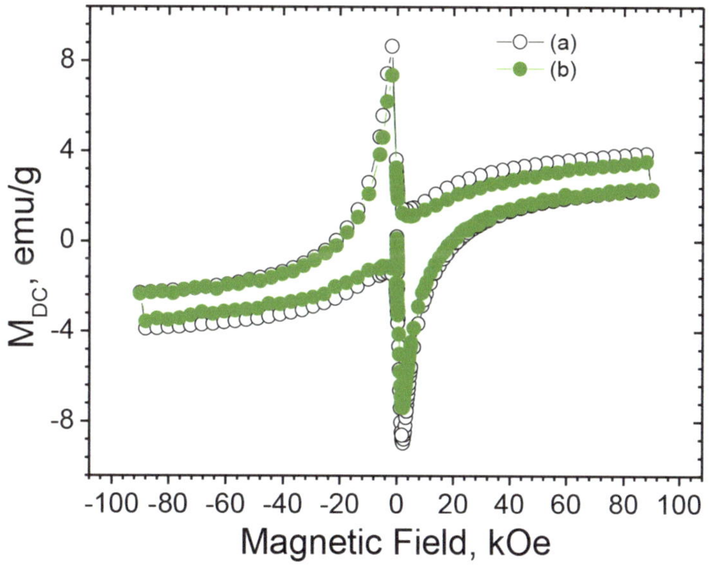


**Figure 2.**
*Magnetic hysteresis loops (magnetization curves) of the obtained composite material before (curve (a)) and after alkaline treatment [16].*

synthesis. It was doped with $Fe_3O_4$ nanopowder, which is imported after the second stage of synthesis. The obtained results show that additions of $Fe_3O_4$ nanopowder at low concentrations to the Re(Y;Dy)-Ba-Cu-O systems improve the superconducting properties by increasing the critical temperature and critical current density of the obtained polycrystalline bulk ceramics with inhomogeneous composition. Homogeneous distribution of iron is observed on their surface, respectively [17].

The multiphase structure is observed in the SEM micrographs of the sample Dy123 with $Fe_3O_4$ shown in **Figure 3**. The main Dy123 phase has a typical surface with elongated grains and an average grain size of ~3.74 μm. The EDX confirms the non-monophasic composition of the ceramic sample as seen in **Table 1**.

In **Figure 4** the EDX mapping of Dy123 with $Fe_3O_4$ sample is shown. Whole crystals of CuO and $BaCuO_2$ are visible on the surface, with small quantities of Fe also detected scattered around the Dy123, CuO and $BaCuO_2$ crystals. We believe that Fe does not react with the other elements and does not form phases of its own.

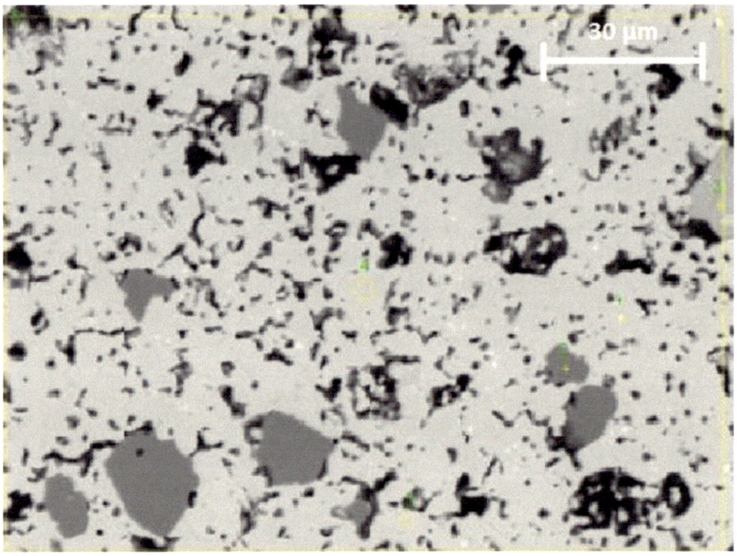


**Figure 3.**
*SEM of Dy123 with Fe3O4 [17].*

| **Spectrum element** | **Series** | **1 Atom. C. [at. %]** | **2 Atom. C. [at. %]** | **3 Atom. C. [at. %]** | **4 Atom. C. [at. %]** | **5 Atom. C. [at. %]** | **6 Atom. C. [at. %]** |
|---|---|---|---|---|---|---|---|
| Dysprosium | L | 11.65 | 0.28 | 0 | 4.68 | 4.71 | 4.30 |
| Barium | L | 9.04 | 0.26 | 12.59 | 9.34 | 9.39 | 8.52 |
| Copper | K | 11.13 | 44.99 | 17.70 | 13.25 | 13.22 | 14.61 |
| Oxygen | K | 68.19 | 54.47 | 57.07 | 71.57 | 71.40 | 71.13 |
| Chlorine | K | 0 | 0 | 12.33 | 0 | 0 | 0.42 |
| Iron | K | 0 | 0 | 0.32 | 1.16 | 1.29 | 1.02 |
| Phase | | DyBaCu | CuO | $BaCuO_2$+ Fe | Dy123+ Fe | Dy123+ Fe | Dy123+ Fe |

**Table 1.**
*Elemental composition of the Dy123 + $Fe_3O_4$ sample obtained by EDX analysis [17].*

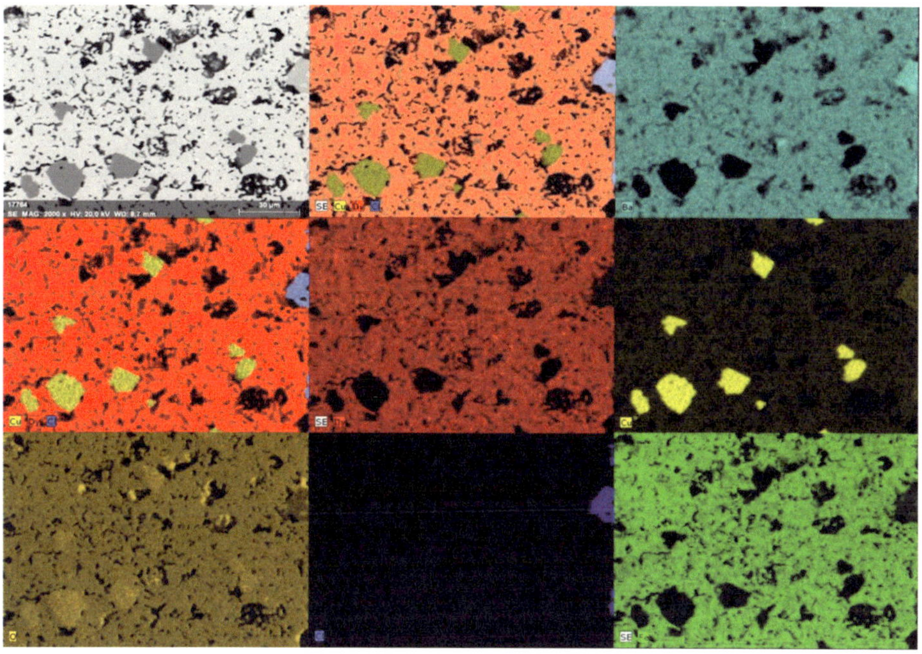

**Figure 4.**
*EDX mapping analysis of Dy123 with $Fe_3O_4$ [17].*

The significant benefits that the rechargeable alkaline Ni-Zn battery system can provide, including its high specific energy and power density, affordable electrode materials, ecologically benign chemistry, and straightforward metal recycling procedure, are what have sparked interest in these batteries [44–48]. The durability of a Ni-Zn battery's Zn anode, which is typically a paste-type composite electrode with a main component of electrochemically active ZnO powder and various additives to enhance the electrode's electrochemical properties, is a major factor in the battery's life cycle. The electrochemical heterogeneity of the anode mass and active surface loss during charge/discharge cycling, which results in progressive changes to the electrode shape and capacity loss, are mostly caused by ZnO's poor electrical conductivity [44, 49]. A powder additive, such as acetylene black, is added to the zinc electrode mass to increase conductivity, although this could cause a rise in hydrogen gas evolution during the charging process. Next, we will go over the study of superconducting BSCCO ceramics as an additive to the Zn electrode mass. It improves the electrode's properties, conductivity, and structural stability, making it a suitable additive to a Zn electrode in alkaline battery systems.

For the first time, structural and morphological changes in B(Pb)SCCO 2212 and B(Pb)SCCO 2201 ceramics were investigated before and after their electrochemical treatment (CV and CP measurements) in a Ni-Zn battery-like medium (7 M KOH, 25°C) [21]. This is the first step toward understanding the observed improvements in the characteristics of the Zn electrode when using ceramic additives. After CV (**Figure 5**), various reduction products are found, such as CuO, $Bi_2O_3$, $Bi_2CuO_4$, $Ca(OH)_2$, and $Sr.(OH)_2$, and the CV curves show that at higher negative potentials the Cu and Bi compounds are further reduced to metallic Cu and Bi (1). This can lead to the formation of metal particles that improve the overall conductivity of the electrode as well as the contact between the particles of the active ZnO. On the other hand, the

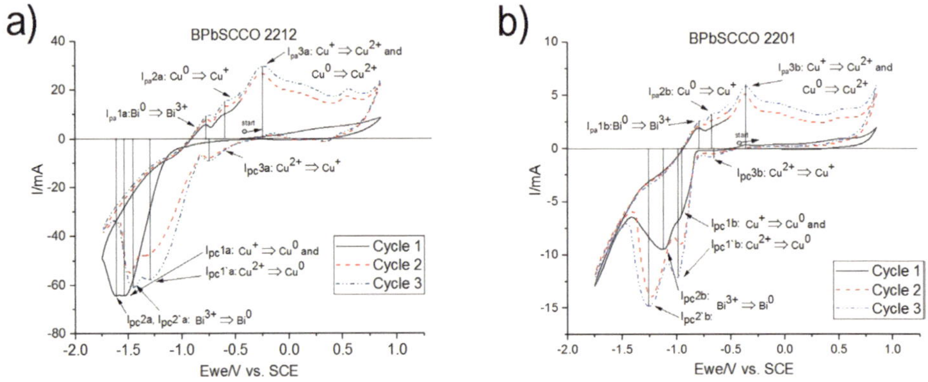


**Figure 5.**
*CV curves of B(Pb)SCCO 2212 (a) and B(Pb)SCCO 2201 (b) in 7M KOH solution. Scan rate: 5 mV $s^{-1}$ [21].*

Ca- and Sr- products in the case of the B(Pb)SCCO 2212 modification do not participate in the redox processes.

The structural and morphological changes occurring in B(Pb)SCCO 2212 and B(Pb)BSCCO 2201 conducting ceramics in an alkaline environment have been investigated, providing additional insights into the mechanisms of the ongoing processes. Electron microscopy reveals that after electrochemical treatment, B(Pb)SCCO 2212 becomes smoother and forms higher and larger facets, while the surface of B(Pb) SCCO 2201 becomes grain-covered [21].

Comparing the X-ray diffraction patterns of B(Pb)SCCO 2212 and B(Pb)SCCO 2201 ceramics before and after their chrono-potentiometric study shows that the primary phases are preserved with reduced intensity after electrochemical treatment, and the appearance of Bi2O3 is observed [21].

The potential of B(Pb)SCCO 2201 and B(Pb)SCCO 2212 undergoes a sharp change when current is applied, indicating an initial breakdown reaction of the ceramic additive before the reduction reactions. The electrode potential remains relatively constant, which suggests that the breakdown is unaffected by the reduction products. The process may continue until the complex structure of B(Pb)SCCO ceramics is completely reduced, leading to the formation of simpler oxides such as CuO, $Bi_2O_3$, $Bi_2CuO_4$, $Ca(OH)_2$, and $Sr(OH)_2$ [21].

These results demonstrate that when the ceramics are treated in an alkaline environment, they undergo partial reduction to oxides and hydroxides, leading to the formation of insoluble Zn compounds, which reduce the solubility of ZnO in the electrolyte and suppress the shape change of the electrode, thereby improving conductivity. In the case of B(Pb)SCCO 2212, the presence of Ca and Sr. products contributes to the stabilization of the electrode, reducing its solubility in the alkaline electrolyte, suppressing dendrite formation, and mitigating gas evolution during operation [21]. These positive effects are believed to be the cause of the observed up to 30% extension in battery life when using a conducting ceramic as an additive in the zinc electrode of the nickel-zinc system [19].

Ultrasound was also used to mix the additives in the preparation of the electrode mass [24]. The micrographs of ZnO with carbon additive, ZnO with carbon and B(Pb)SCCO 2201 additive, and ZnO with carbon and B(Pb)SCCO 2212 additive obtained by the ultrasound-assisted mixing are shown (**Figure 6**). The ultrasound-assisted mixing of the active mass leads to superior particle distribution, as seen on the SEM images. This is most noticeable in **Figure 6b**. The SEM micrographs of the electrodes provided evidence for their better homogenization, while the impedance measurements confirmed the better conductivity of the zinc mass obtained through ultrasonic irradiation [24, 25]. In **Table 2**, the average values of EDX results are presented. It is seen that the composition did not change during the ultrasound mixing.

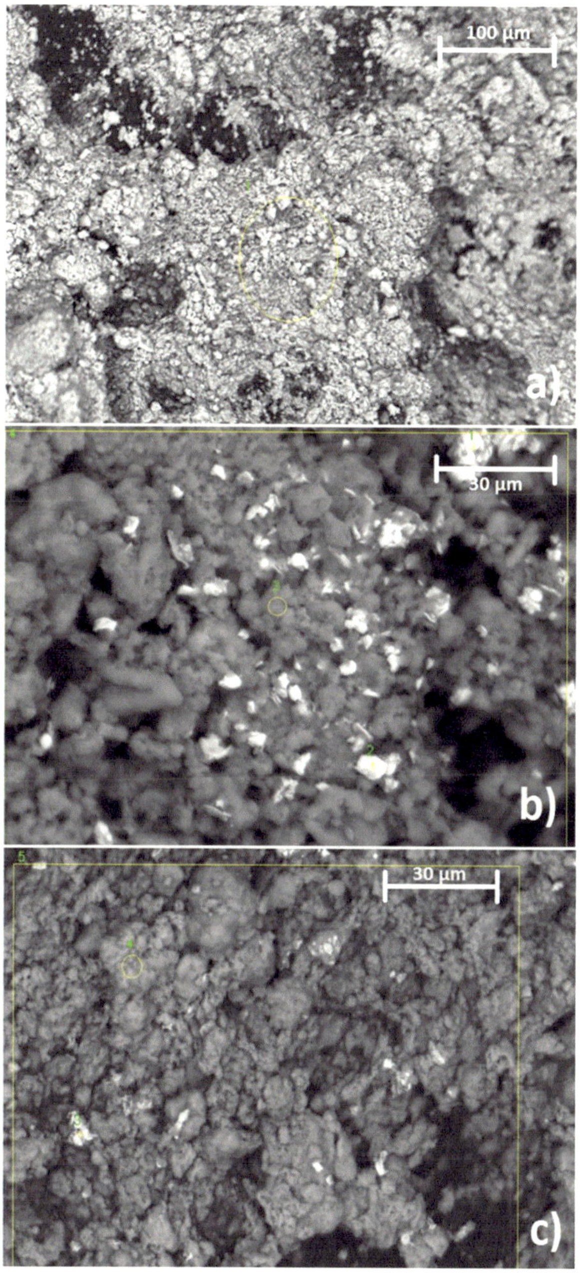


**Figure 6.**
*SEM images of ZnO with carbon additive (a), ZnO with carbon and B(Pb)SCCO 2201 additive (b), and ZnO with carbon and B(Pb)SCCO 2212 additive (c) with EDX analysis [24].*

| Samples | Elemental content, [at. %] | | | | | | | |
|---|---|---|---|---|---|---|---|---|
| | Zn | C | Bi | Pb | Sr | Ca | Cu | O |
| Sample a) | 18.27 | 21.81 | | | | | | 59.91 |
| Sample b) | | | | | | | | |
| Spectrum 1 | 4.25 | | 6.57 | 1.34 | 7.04 | | 3.74 | 77.07 |
| Spectrum 2 | 25.33 | | | | | | 0.36 | 74.31 |
| Sample c) | | | | | | | | |
| Spectrum 3 | 5.2 | | 6.41 | 0.76 | 6.19 | 5.35 | 7.57 | 68.52 |
| Spectrum 4 | 46.76 | | | | | | | 53.24 |

**Table 2.**
*EDX analysis results for ZnO with carbon additive (Sample a), ZnO with carbon and 2201 additive, and ZnO (Sample b) with carbon and 2212 additive (Sample c) [24].*

During the ultrasound treatment, the composition of the active mass is preserved. Physically mixing the B(Pb)SCCO additives with active zinc mass for Ni-Zn battery has already been proven to have an influence in our past investigations. The PEIS measurements show that the overall resistance of the electrode and that of the charge transfer in the electric double layer is reduced by the BSCCO additives and thus the electrochemical reaction is facilitated. These effects are strengthened due to the ultrasonic treatments, although to a very low degree. Due to that, we believe that better homogenization through ultrasonic irradiation of the additives with the ZnO may enhance the conductivity, the stability and increase the life of the battery even more. This study gives us a basis for future investigation of the newly prepared Zn electrodes in a Ni-Zn electrochemical cell [25].

In more detail, the structural, morphological, and electrochemical changes that might occur in the active mass and the electrodes, due to the different methods of preparation, have been examined: ball-milling treatment, ultrasound, and mechanical (by hand) mixing [25]. This study has proven useful in explaining how the different methods of preparation might improve or worsen the homogenization of the zinc mass content and the overall performance of the Zn-electrode. The results show that the method of grinding in a ball-mill does not contribute to better homogenization of the ceramic additives in the active zinc mass. The ultrasound treatment for ½ hour leads to better homogenization of the active mass with B(Pb)SCCO 2212 and B(Pb)SCO 2201 ceramic and respectively the reduction of the electrode resistance. The reason for the observed results can be explained on one hand by the processes that take place during ultrasonic treatment and on the other hand by the presence of calcium carbonate in B(Pb)SCCO 2212, which probably favors faster processes at the micro level. In the absence of B(Pb)SCCO 2212 more time is needed for the manifestation of the desired effect [25].

The impedance characteristics of electrochemical systems (ECSs) were also investigated, in which the anodes are obtained from nanosized ZnO material doped with high-temperature superconducting ceramic $Bi_{1.7}Pb_{0.3}Sr_2Ca_2Cu_3O_y$ (B(Pb)SCCO 2212) at a concentration of 5, 7 or 10 wt.% for the purpose of electrochemical applications (such as Ni-Zn batteries) [26]. The electrically conductive properties of such alkaline electrolyte ECSs were characterized at ambient temperature by electrochemical impedance spectroscopy measurements. The frequency spectra of $Z^*$ of the produced ECSs were recorded by electrical impedance-meter Bio-Logic SP–200, in the range 0.1 Hz – 100 kHz, at room temperature (in our case 25°C). By this technique [50, 51], both real (*ReZ*) and imaginary (*ImZ*) parts of the complex electrical impedance $Z^* = ReZ + i\ ImZ$ are simultaneously measured as a function of the frequency *f* of the alternating-current (AC) electric field applied. The amplitude of the AC voltage applied between the electrodes of ECSs was 10 $mV_{RMS}$ (sine function). The dimensions of the electrodes were 5.0 cm × 3.0 cm, their thicknesses were 0.15 cm, and the distance between them was 36 mm. They were separated by a microporous separator in an alkaline electrolyte (liquid KOH). For the sake of comparison, reference ECSs with identical geometry were designed with copper (Cu) (99.99% pure) or undoped ZnO anodes [26].

The effect of ceramic additives on the complex electrical impedance and electrical conductivity of the studied ECSs was evaluated. It was demonstrated that the incorporation of B(Pb)SCCO 2212 HTSC ceramics at a concentration of 7 wt% into the ZnO anode material of ECSs leads to an increase in their static (DC) electrical conductivity. In addition, the AC conductivity of the ECSs is also improved and approaches the values corresponding to ECSs with undoped ZnO anode through identical ECSs geometrical configurations. The Nyquist plots in **Figure 7a** clearly show the higher electrical conductivity of the studied ECSs having ZnO anode doped with BPSCCO ceramics, as compared to the undoped ZnO anode.

Also, it is apparent in **Figure 7a** that $R_B$ of the active volume of the studied ECSs is decreased when the BPSCCO percentage in the ZnO mass of the anode of the ECSs was increased from 5 wt% to 7 wt%. By further increase of the BPSCCO concentration up to 10 wt%, $R_B$ slightly diminishes, and practically there is no difference in resistance between identically constructed ECSs with the anode of ZnO doped with the BPSCCO at either 7 wt% or 10 wt% (**Figure 7a**). The same applies for the corresponding values of the electrical conductivity $\sigma = d/(R_B A)$. Note also that at these two concentrations (7 wt% and 10 wt%) of BPSCCO, $ImZ < ReZ/2$ for the maximum *ImZ* value of the circles in the Nyquist plots, i.e., the dispersive character of capacity is evident, in contrast to the case for 5 wt% BPSCCO. Accordingly, the ECS scheme

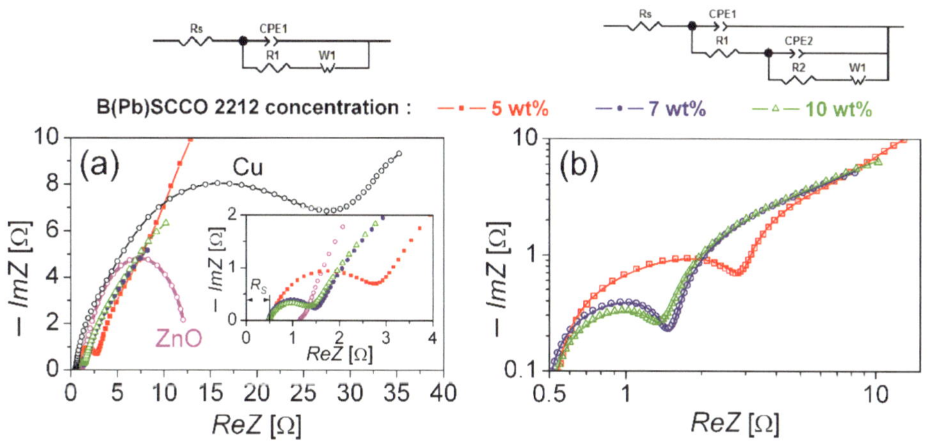


**Figure 7.**
*(a) Nyquist plots for complex impedances measured for the studied ECSs. The designations are as in* **Figure 1***. The insert: Enlarged view; (b) fitted data for ECSs with BPSCCO-doped ZnO anodes (data from (a)) to the equivalent circuit scheme shown at the top of (b). The lines represent the fitting results [26].*

needs to be modified by adding new elements, as drawn in **Figure 7b**. Such an ECS model gives satisfying fit results that agree very well with the experimental data points (**Figure 7b**).

The electrical properties of the alkaline systems whose Zn electrodes (anodes) contain active mass produced from composites of ZnO and B(Pb)SCCO 2212 conductive cuprate ceramics (5, 7 and 10 wt%) were investigated in terms of complex impedance (Z*) and tangent loss spectra [27]. The aim was to trace the effect of the ceramics as additives on the impedimetric response and tangent loss spectra of the electrochemical cells. The analysis of the electric impedance spectra of such electrochemical systems indicated that the redox processes in them are enhanced by the increase of the concentration of the B(Pb)SCCO 2212 ceramic additives (in the range 5–10 wt.%). Significantly, at a lower concentration of B(Pb)SCCO 2212, for example, 5 wt.%, a lower electric loss was established for the examined cells (related to the better surface properties of the layer deposited on the anode). Thus, such composite material (Zn active mass) with included B(Pb)SCCO 2212 conductive ceramics can be proper for producing electrodes in Ni-Zn electrochemical alkaline systems with enhanced performance.

For the first time, copper-based superconducting ceramics have been proposed as additives to the negative electrode of nickel/zinc batteries. The addition of copper-based superconducting ceramics such as BSCCO (2201, 2212, 2223) or YBCO (123) improves the bulk conductivity and structure of the zinc electrode, reduces gas evolution during electrode charging, which has a favorable effect on capacity retention and extends the electrode's lifespan (**Figure 8**) [19].

These positive effects of the B(Pb)SCCO ceramic, observed from the tested electrochemical systems, make it a promising doping material for the development of new electrochemical systems as well as for use in different devices.

The dielectric properties of the samples in the system $Bi_2O_3$-$TiO_2$-$Nd_2O_3$ have been studied in frequency 2.7 GHz and control of dielectric parameters is realized by precise control of the percentage of initial oxides and synthesis temperatures [40]. Depending on the controlled melting conditions and additional heat treatment of the

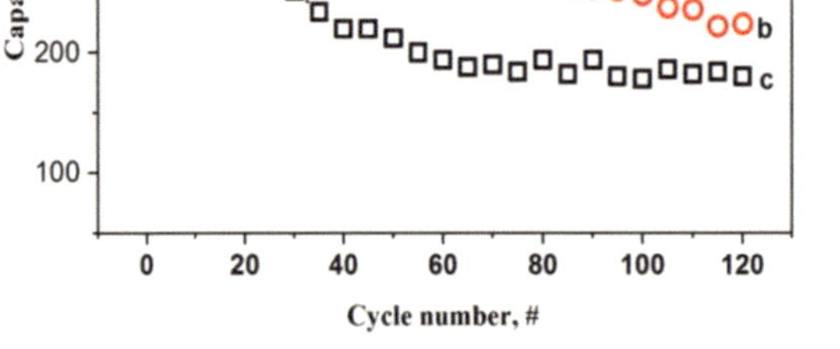


**Figure 8.**
*Dependency of capacity on the number of cycles for a nickel-zinc alkaline battery cell with zinc electrodes containing additives of BSCCO 2212 copper-based superconducting ceramics (curve a), YBCO 123 copper-based superconducting ceramics (curve b), and carbon material - acetylene black (curve c) [19].*

supercooled compositions, different polyphase glass-ceramic materials with different microstructures in the systems $Bi_2O_3$-$TiO_2$-$SiO_2$ and $Bi_2O_3$-$TiO_2$-$Nd_2O_3$ controlled microstructural features and dielectric properties were obtained [52].

The obtained results are the built and patented systems of new materials based on ferroelectric ceramics, with application for capacitor batteries with specific thermal parameters [53, 54]. In addition to that, a new sequence of technological procedures has been created, which relates to the production of conductive composite ceramic materials for industrial use [53]. Furthermore, conductive ceramics have been made for use in hydrogen generators [54].

## 4. Conclusions

In summary, the scientific and applied contributions can be clarified as:

Creation of new composite materials from natural raw materials and methods for their industrial production.

A composite material has been obtained, consisting of the Y123 phase and the $BaCuO_2$ phase in a ratio of 2:1. It exhibits superconductivity at temperatures higher than that of liquid nitrogen, displays magnetic properties, and is resistant in alkaline environments. It finds applications in the field of electrical industry, computer technology, space technology, medicine, alkaline rechargeable batteries used as power sources, and more.

New constructions and materials for ceramic capacitor batteries.

The composition of the active mass of the negative zinc electrode, consisting of powdered zinc and/or zinc oxide and binders with high hydrogen evolution overpotential, also includes an additive of copper superconductive ceramic. It improves the volumetric conductivity and structure of the zinc electrode, reduces gas evolution during electrode charging, which has a favorable effect on capacity stability and extends the electrode's lifespan, and can find application as a negative electrode in alkaline rechargeable batteries, particularly nickel-zinc batteries.

## Acknowledgements

The works presented in this chapter are developed as part of contract №: BG-RRP-2.004-0002-C01, project name: BiOrgaMCT, Procedure BG-RRP-2.004 "Establishing of a network of research higher education institutions in Bulgaria", funded by BULGARIAN NATIONAL RECOVERY AND RESILIENCE PLAN and is part of an inter-academic collaboration project between the Bulgarian Academy of Sciences, the Estonian Academy of Sciences, Tallinn University of Technology and the Institute of Low Temperature and Structure Research, Polish Academy of Sciences.

Dedicated to our teacher prof. Yanko Dimitriev.

We express our gratitude to our colleagues, who participated in the realization of these scientific developments, and to academician Alexander Petrov for his invaluable advice and support.

## Author details

Angelina Stoyanova-Ivanova[1*] and Stanislav Slavov[2]

1 G. Nadjakov Institute of Solid State Physics, Bulgarian Academy of Sciences, Sofia, Bulgaria

2 Department of Mathematics, University of Chemical Technology and Metallurgy, Sofia, Bulgaria

*Address all correspondence to: angelina@issp.bas.bg

## References

[1] Liu HJ, Luo MQ, Yang LX, Zeng CL, Fu C. A high strength and conductivity bulk MAGNÉLI phase $Ti_4O_7$ with superior electrochemical performance. Ceramics International. 2022;**48**(17):25538-25546. DOI: 10.1016/j.ceramint.2022.05.233

[2] Chen M, Liu Q, Zhu J, Cao W, Chen J, Su J. The influence of $ZrO_2$ doping on microstructure and electrical properties of zno-based conductive ceramics. Ceramics International. 2021;**47**(17):24959-24965. DOI: 10.1016/j.ceramint.2021.05.223

[3] Bondarchuk AN, Corrales-Mendoza I, Tomás SA, Marken F. A hematite photoelectrode grown on porous and conductive $SNO_2$ ceramics for solar-driven water splitting. International Journal of Hydrogen Energy. 2019;**44**(36):19667-19675. DOI: 10.1016/j.ijhydene.2019.06.055

[4] Wang G, Liu Y, Duan Y, Ye J, Lin Z. Effects of porosity on the electrochemical oxidation performance of $Ti_4O_7$ electrode materials. Ceramics International. 2023;**49**(10):15357-15364. DOI: 10.1016/j.ceramint.2023.01.120

[5] Wu MK, Ashburn JR, Torng CJ, Hor PH, Meng RL, Gao L, et al. Superconductivity at 93 K in a new mixed-phase Y-ba-cu-O compound system at ambient pressure. Physical Review Letters. 1987;**58**(9):908-910. DOI: 10.1103/physrevlett.58.908

[6] Andreouli C, Tsetsekou A. Processing effect on microstructure and superconducting properties of sintered $ReBa_2Cu_3Oy$ ceramics — The role of ionic radius. Journal of the European Ceramic Society. 2000;**20**(12):2101-2114. DOI: 10.1016/s0955-2219(00)00074-1

[7] Karppinen M, Yamauchi H. Hole-doping routes for understanding the relationship between atomic arrangements and superconductivity properties in multi-layered copper oxides. International Journal of Inorganic Materials. 2000;**2**(6):589-599. DOI: 10.1016/s1466-6049(00)00085-4

[8] Marsh P, Fleming RM, Mandich ML, DeSantolo AM, Kwo J, Hong M, et al. Crystal structure of the 80 K superconductor YBa2Cu4O8. Nature. 1988;**334**(6178):141-143. DOI: 10.1038/334141a0

[9] Karpinski J, Rusiecki S, Bucher B, Kaldis E, Jilek E. The nonstoichiometry of the high-Tc superconductor Y2Ba4Cu7O15±x (14 K⩽Tc⩽68 K). Physica C: Superconductivity. 1989;**161**(5-6):618-625. DOI: 10.1016/0921-4534(89)90398-5

[10] Genoud J-Y, Graf T, Triscone G, Junod A, Muller J. Variation of the superconducting and structural properties of $Y_2Ba_4Cu_7Oz$ with oxygen content (14.1 < z < 15.3, 30 K ⩽ Tc ⩾ 95 K). Physica C: Superconductivity. 1992;**192**(1-2):137-146. DOI: 10.1016/0921-4534(92)90753-y

[11] Aliabadi A, Akhavan Farshchi Y, Akhavan M. A new Y-based HTSC with Tc above 100K. Physica C: Superconductivity and Its Applications. 2009;**469**(22):2012-2014. DOI: 10.1016/j.physc.2009.09.003

[12] Superconductors. Superconductors. Available from: http://www.superconductors.org/

[13] Sreedhar K, Ganguly P. Magnetic susceptibility studies on ternary oxides of copper(II). Inorganic Chemistry. 1988;**27**(13):2261-2269. DOI: 10.1021/ic00286a011

[14] Ganguly P, Sreedhar K, Raju AR, Demazeau G, Hagenmuller P. Electron paramagnetic resonance studies of some ternary oxides of copper (II). Journal of Physics: Condensed Matter. 1989;**1**(1):213-226. DOI: 10.1088/0953-8984/1/1/018

[15] Stoyanova-Ivanova A, Terzieva S, Georgieva S, Blagoev B, Kovacheva D, Zaleski A, et al. Superconductivity and magnetic studies of bulk Y123/$BaCuO_2$ composite. Romanian Journal of Physics. 2018;**61**:1-2

[16] Stoyanova-Ivanova A, Terzieva S, Blagoev B, Georgieva-Kiskinova S, Kovacheva D. Composite material and method of obtaining it. 2020. Patent #67066

[17] Stoyanova-Ivanova A, Kolev S, Petrova V, Petkov O, Tran L, Babij M, et al. Physicochemical study of bulk Dy-123 doped with nano-$Fe_3O_4$. Bulgarian Chemical Communications. 2022;**54**(Special Issue B1):71-76. DOI: 10.34049/bcc.54.B1.0404

[18] Stoyanova-Ivanova A, Terzieva DS, Ivanova DG, Mladenov AM, Kovacheva GD, Raicheff GR, et al. The use of high-temperature superconducting cuprate as a dopant to the negative electrode in Ni-Zn batteries. Bulgarian Chemical Communications. 2015;**47**(1):221-228

[19] Mladenov M, Raicheff R, Stoyanov L, Stoyanova-Ivanova A, Terzieva S, Kovacheva D. Electrode mass for zinc electrode of alkaline rechargeable batteries. 2018. Patent #66730

[20] Lilov PA, Vasilev AY, Stoyanova AE, Marinov YG, Stoyanova-Ivanova AK. Electrochemical impedance study of HTSC ceramics YBCO and BSCCO in presence of electrolyte. Bulgarian Chemical Communications. 2018;**50**(Special issue A):153-157

[21] Stoyanova-Ivanova A, Lilov P, Vasev A, Stoyanova A, Ivanova G, Karashanova D, et al. Studies of structural and morphological properties of cuprate conductive ceramics after electrochemical treatment in alkaline electrolyte. Materials Chemistry and Physics. 2020;**239**:121934. DOI: 10.1016/j.matchemphys.2019.121934

[22] Stoyanova-Ivanova A, Vasev A, Lilov P, Petrova V, Marinov Y, Stoyanova A, et al. Conductive ceramic based on the Bi-Sr-Ca-Cu-O HTSC system as an additive to the zinc electrode mass in the rechargeable Ni-Zn batteries – Electrochemical impedance study. Proceedings of the Bulgarian Academy of Sciences. 2019;**72**(2):174-171. DOI: 10.7546/crabs.2019.02.05

[23] Ivanova G, Stoyanova-Ivanova A, Kovacheva D, Stoyanova A. Improvement of the electrochemical properties of Ni-Zn rechargeable batteries by adding B(Pb)SrCaCuO conducting ceramics. Bulgarian Chemical Communications. 2019;**51**(1):66-72

[24] Petrova V, Stoyanova-Ivanova A, Lilov P, Petkov O, Ivanova G, Karamanova B, et al. Ultrasound assisted mixing of zinc active mass with conductive ceramic additives for Ni-Zn battery. ECS Transactions. 2019;**95**(1):227-234. DOI: 10.1149/09501.0227ecst

[25] Stoyanova-Ivanova A, Ivanova G, Leffterova E, Petrova V, Petkov O, Stoyanova A. Influence of various preparation methods on the electrochemical properties of the active Zn mass with conductive cuprate ceramics as additives. Current Topics in Electrochemistry. 2021;**23**:25-32

[26] Stoyanova-Ivanova A, Lefterova E, Ivanova G, Marinov Y, Hadjichristov G, Stoyanova A. Impedance spectroscopy study of Ni-Zn electrochemical alkaline

systems with anode of nano-sized zno doped with conducting ceramics b(pb) scco 2212. Proceedings of the Bulgarian Academy of Sciences. 2022;**75**(3):358-366. DOI: 10.7546/crabs.2022.03.05

[27] Angelina, Lefterova E, Marinov Y, Ivanova G, Stoyanova A, Hadjichristov G. Electrochemical cells with anodes performed with Zn active mass with added B(Pb)SCCO 2212 conductive ceramics. Comptes rendus de l'Académie bulgare des Sciences. 2023;**76**(6):871-880

[28] Buhay H, Sinharoy S, Kasner WH, Francombe MH, Lampe DR, Stepke E. Pulsed laser deposition and ferroelectric characterization of bismuth titanate films. Applied Physics Letters. 1991;**58**(14):1470-1472. DOI: 10.1063/1.105200

[29] Maeder MD, Damjanovic D, Voisard C, Setter N. Piezoelectric properties of Srbi4ti$_4$O$_{15}$ ferroelectric ceramics. Journal of Materials Research. 2002;**17**(6):1376-1384. DOI: 10.1557/jmr.2002.0205

[30] Wang C-M, Wang J-F. Aurivillius phase potassium bismuth titanate: $K_{0.5}Bi_{4.5}Ti_4O_{15}$. Journal of the American Ceramic Society. 2008;**91**(3):918-923. DOI: 10.1111/j.1551-2916.2007.02211.x

[31] Saito Y, Takao H, Tani T, Nonoyama T, Takatori K, Homma T, et al. Lead-free piezoceramics. Nature. 2004;**432**(7013):84-87. DOI: 10.1038/nature03028

[32] Takenaka T, Nagata H, Hiruma Y. Current developments and prospective of lead-free piezoelectric ceramics. Japanese Journal of Applied Physics. 2008;**47**(5):3787-3801. DOI: 10.1143/jjap.47.3787

[33] Zhang S-T, Kounga AB, Aulbach E, Ehrenberg H, Rödel J. Giant strain in lead-free piezoceramics $Bi_{0.5}Na_{0.5}TiO_3$–$BaTiO_3$–$K_{0.5}Na_{0.5}NbO_3$ system. Applied Physics Letters. 2007;**91**(11):112906. DOI: 10.1063/1.2783200

[34] N. F. Borrelli, Andrew Herczog, R. D. Maurer, Electro-optic effect of ferroelectric microcrystals in a glass matrix, Applied Physics Letters 7, 117-118 (1965), doi.org/10.1063/1.1754333

[35] Bell AJ. Ferroelectrics: The role of ceramic science and engineering. Journal of the European Ceramic Society. 2008;**28**(7):1307-1317. DOI: 10.1016/j.jeurceramsoc.2007.12.014

[36] Graça MP, Ferreira da Silva MG, Valente MA. Structural and electrical properties of $SiO_2$–$Li_2O$–$NB_2O_5$ glass and glass-ceramics obtained by thermoelectric treatments. Journal of Materials Science. 2007;**42**(8):[25]43-2550. DOI: 10.1007/s10853-006-1208-z

[37] Shankar MV, Varma KBR. Crystallization of ferroelectric bismuth vanadate in $Bi_2O_3$–$V_2O_5$–$SRB_4O_7$ glasses. Journal of Non-Crystalline Solids. 1998;**226**(1-2):145-154. DOI: 10.1016/s0022-3093(97)00490-0

[38] Pengpat K, Holland D. Glass-ceramics containing ferroelectric bismuth germanate (BI2GEO5). Journal of the European Ceramic Society. 2003;**23**(10):1599-1607. DOI: 10.1016/s0955-2219(02)00410-7

[39] Koutzarova T, Nedkov I, Ausloos M, Cloots R, Midlarz T, Nogues M. The influence of the polycrystalline state and partial Dy substitution on the superconducting properties of YBCO. Physica Status Solidi (A). 2002;**191**(1):235-242. DOI: 10.1002/1521-396x

[40] Slavov SS, Soreto Teixeira S, Graça MP, Costa LC, Popova V,

Dimitriev YB. $Bi_2O_3$-$TiO_2$-$Nd_2O_3$lead-free material for microwave device applications. International Journal of Applied Glass Science. 2018;**10**(2):202-207. DOI: 10.1111/ijag.12976

[41] Afify A, Slavov S, Mahmoud A, Hassan M, Ataalla M, Staneva A, et al. Determination of the sensing characteristics of $SiO_2$-$Bi_2O_3$-$TiO_2$ system toward relative humidity. Journal of Chemical Technology and Metallurgy. 2018;**53**(6):1073-1080

[42] Raykov R, Staneva A, Dimitriev Y, Slavov S, Soreto Teixeira S, Prezas PR, et al. Dielectric relaxation in glass and glass-ceramic materials of the system $La_2O_3$-$Gd_2O_3$-pbo-mno-$B_2O_3$. International Journal of Applied Glass Science. 2018;**10**(1):75-82. DOI: 10.1111/ijag.12553

[43] Nedeltcheva T, Georgieva S, Vladimirova L, Stoyanova-Ivanova A. Increasing the sensitivity of the spectrophotometric determinations of the oxygen content in YBCO superconducting samples using the I3−-starch compound. Talanta. 2009;**77**(5):1745-1747. DOI: 10.1016/j.talanta.2008.10.019

[44] Linden D, Reddy TB. Chapter 31: Nickel-zinc batteries. In: Handbook of Batteries. 3rd ed. New York, New York, USA: McGraw Hill; 2004

[45] Shukla A. Nickel-based rechargeable batteries. Journal of Power Sources. 2001;**100**(1-2):125-148. DOI: 10.1016/s0378-7753(01)00890-4

[46] McLarnon FR, Cairns EJ. The secondary alkaline zinc electrode. Journal of the Electrochemical Society. 1991;**138**(2):645-656. DOI: 10.1149/1.2085653

[47] Geng M. Development of advanced rechargeable Ni/MH and Ni/Zn batteries. International Journal of Hydrogen Energy. 2003;**28**(6):633-636. DOI: 10.1016/s0360-3199(02)00137-4

[48] Phillips J, Mohanta S, Geng M, Barton J, McKinney B, Wu J. Environmentally friendly nickel-zinc battery for high rate application with higher specific energy. ECS Transactions. 2009;**16**(16):11-17. DOI: 10.1149/1.3087437

[49] Yuan YF, Yu LQ, Wu HM, Yang JL, Chen YB, Guo SY, et al. Electrochemical performances of Bi based compound film-coated ZnO as anodic materials of Ni–Zn secondary batteries. Electrochimica Acta. 2011;**56**(11):4378-4383. DOI: 10.1016/j.electacta.2011.01.006

[50] Barsoukov E, Macdonald JR, editors. Impedance Spectroscopy: Theory, Experiment, and Applications. Hoboken, NJ, USA: John Wiley & Sons; 2005. DOI: 10.1604/9780471647492

[51] Lvovich VF. Impedance Spectroscopy: Applications to Electrochemical and Dielectric Phenomena. Hoboken, NJ, USA: John Wiley & Sons; 2012

[52] Slavov S, Krapchanska M, Kashchieva E, Parvanov S, Dimitriev Y. Synthesis of doped bismuth titanate ceramics with $Nd_2O_3$ and $SiO_2$ and their electrical properties. Journal of Chemical Technology and Metallurgy. 2013;**48**(2):174-178

[53] Petar NG, Venetka SP, Stanislav SS, Andrei ID. Composition of electrically conductive composite ceramic material and method for its preparation. Pending patent №RO1358. U/00020/26.05/2020. 2020.

[54] Lakov L, Slavov S, Toncheva K, Dimitriev Y. Ceramic battery. Patent № BG67056/01.06. 2020

Section 3

# Functional Studies Based on Traditional Ceramics

Chapter 7

# Investigation of Grinding Force and Process Optimization of Woven Ceramic Matrix Composites

*Bin Lin, Jingguo Zhou, Jinhua Wei, Haoji Wang and Tianyi Sui*

## Abstract

Due to the special structure of the woven ceramic matrix composites, challenges and difficulties in the grinding process gradually emerge. How to ensure the reasonable and reliable application of materials has become a hot topic in the present research. The force model is beneficial to understand, predict, and even control the machining process. This chapter investigates the grinding force and process optimization of woven ceramic matrix composites, especially grinding force modeling, surface quality, and process optimization of woven ceramic matrix composites during grinding. A new force model considering the fiber orientation of WCMC is developed based on the energy balancing theory. Through the construction of a mathematical model, the study demonstrates the correlation of grinding force with the processing parameters and the composite fiber orientation. The optimum process parameters were obtained by aiming at minimum grinding force and maximum surface quality. The results show that the predictable model has good consistency with the experimental results, and fiber orientation has a major influence on the grinding force. This research can be used to predict the grinding force, thus conducting the machining and controlling their processing quality.

**Keywords:** woven ceramic matrix composites, grinding, process optimization, force model, surface quality

## 1. Introduction

Woven ceramic matrix composites (WCMC) are more and more widely used in the military, transportation [1], shipping, aerospace, nuclear industry [2–4], and other fields due to their excellent properties of high specific stiffness, high-temperature resistance, corrosion resistance, high toughness, and wear resistance. However, due to the anisotropy of its materials, it is difficult to carry out mechanical processing [5]. Faced with the stability, precision, and low damage requirements of braided ceramic matrix composites, the most common machining method for WCMC is grinding [6, 7].

Grinding force is an important index to characterize the influence of grinding parameters on the grinding process. Grinding force is produced by elastic deformation, plastic deformation of the workpiece and the interaction between abrasive particle, workpiece and grinding chip [8–10]. It is related to almost all process factors of grinding. Therefore, it is of great significance to establish a grinding force prediction model [11]. For the grinding of composite materials, research on the machining force of ceramic material both on experimental and analytical aspects has been in-depth investigated [12–15]. Durgumahanti et al. [16] established a new grinding force model developed by incorporating the effects of variable coefficient of friction and plowing force. Li et al. [17] developed a new grinding force model for micro-grinding of reaction-bonded silicon carbide (RB-SiC) ceramics. It was found that plowing and fracture were the dominant removal modes. Furthermore, comparisons are made between the values predicted and the experimental data, indicating that the proposed model is acceptable and can be used to simulate the grinding force for RB-SiC ceramics in practice. Zhang et al. [12] studied the grinding force of crystalline ceramic materials and amorphous ceramic materials in the experiment to improve the mechanical models of ceramics. The result shows that the grinding force models are not applicable for non-crystalline ceramic fused silica and the specific grinding energy fluctuates irregularly as a function of the maximum undeformed chip thickness seen from the experiment. So far, one of the main focuses is on ultrasonic-assisted machining [18–20] of WCMC and the corresponding force model to improve processing quality. But the efficiency is lower and more costly than ordinary machining methods. Though the machining force model of ordinary milling [13, 21], drilling [22], and even grinding [23, 24] of WCMC has also been researched, the theory is always based on fracture mechanics, the maximum undeformed chip thickness, or aggregate force including plowing, scratching, and removing neglecting fiber orientation and idealizing the model because of the complication of woven structure and reinforced particles or short fibers. In the review of the previous work, though WCMC grinding force has been paid great attention to and many experimental studies have already been made, the model considering fiber orientation is a gap. Hence, a grinding force model for WCMC involving fiber orientation is essential.

This chapter mainly develops a grinding for the model of WCMC considering fiber orientation based on the energy balancing theory to fill the gap. The influence of grinding parameters on specific energy and grinding force is analyzed. Taking the $SiO_2f/SiO_2$ as an example, a series of experiments were carried out to verify the accuracy of the model. This research, on one hand, can be used to predict the grinding force of WCMC; on the other hand, it provides a baseline for selecting the proper machine and tool for WCMC processing.

## 2. Modeling of the grinding force of woven ceramic matrix composites

### 2.1 Force modeling of the single particle in grinding WCMC

Points A and C are the positions where the diamond first cuts in and leaves the workpiece, respectively. Points B and D are the locations where particles touch and leave the workpiece for a second time [25]. It is worth noting that point B is the coincident point of the grit motion track, as shown in **Figure 1**. Therefore, the shaded area BCD can be viewed as the amount of particle removal except for the first one (**Figure 1(a)**). Assuming the process is ideal, the material is completely removed in each cut, and the first cut's difference is neglected.

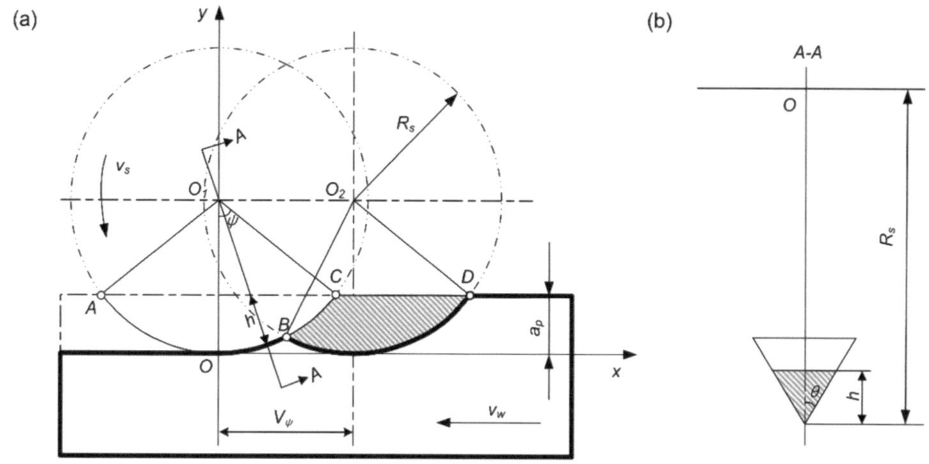


**Figure 1.**
*Particle trajectory during the grinding process: (a) relative motion trajectories of adjacent particles and (b) cross section of the chip thickness [25].*

### *2.1.1 Kinematic analysis of abrasive particles*

In the XOY coordinate system, the grit motion track equation is (**Figure 1(a)**):

$$x = R_s \sin\psi + V_\psi \tag{1}$$

$$y = R_s(1 - \cos\psi) \tag{2}$$

where $R_s$ (mm) is the radius of the grinding wheel, $\psi$ (rad) is the angle between the instantaneous position of grain and vertical direction in the grinding process, and $V_\psi$ (mm) is the horizontal travel distance when the wheel is turned at an angle $\psi$.

Introducing the time variable, the grit relative motion track equation changing over time is

$$x(t) = R_s \sin(v_s t/R_s) + v_w t \tag{3}$$

$$y(t) = R_s(1 - \cos(v_s t/R_s)) \tag{4}$$

### *2.1.2 Cross-sectional area of the chip thickness*

According to **Figure 1(a)**, it can be obtained as follows:

$$\cos(v_s t/R_s) = (R_s - a_p)/(R_s - h) \tag{5}$$

$$h = R_s - (R_s - a_p)/\cos(v_s t/R_s) \tag{6}$$

In **Figure 1(b)**, the cross-sectional area of the chip thickness is

$$s(t) = \frac{1}{2} \cdot 2h\tan\theta \cdot h = h^2 \tan\theta = \left(R_s - \frac{R_s - a_p}{\cos(v_s t/R_s)}\right)^2 \tan\theta \tag{7}$$

where $\theta$ (°) is half of the grain vertex angle. h(mm) is the cutting height of the grain, and $a_p$ (mm) is the grinding depth. Because $\psi = v_s t/R_s$ is so small that $\cos(v_s t/R_s) \approx 1$, then

$$s(t) \approx a_p{}^2 \tan\theta \tag{8}$$

### 2.1.3 Moments grit cutting into and out of workpiece

When grit *cuts* out of the workpiece, $h = 0$ in Eq. (5), so:

$$\cos(v_s t_{\text{nmax}}/R_s) = 1 - a_p/R_s \tag{9}$$

Where $t_{nmax}(s)$ is the *moment* grit cuts out of workpiece for each cutting process. Hence:

$$t_{\text{nmax}} = \frac{R_s}{v_s}\arccos\left(1 - \frac{a_p}{R_s}\right) + 2n\pi\frac{R_s}{v_s} \quad (n = 0, 1, 2, 3\ldots\ldots) \tag{10}$$

When $n = 0$ and $n = 1$, then:

$$t_{0\,\text{max}} = \frac{R_s}{v_s}\arccos\left(1 - \frac{a_p}{R_s}\right), t_{1\,\text{max}} = \frac{R_s}{v_s}\arccos\left(1 - \frac{a_p}{R_s}\right) + 2\pi\frac{R_s}{v_s} \tag{11}$$

It is assumed that the moment grit reaches point $B$ at adjacent two times $t_1(s)$ and $t_2(s)$, so $x(t_1) = x(t_2)$ $y(t_1) = y(t_2)$, according to Eqs. (3) and (4):

$$R_s \sin(v_s t_1/R_s) + v_w t_1 = R_s \sin(v_s t_2/R_s) + v_w t_2 \tag{12}$$

$$R_s[1 - \cos(v_s t_1/R_s)] = R_s[1 - \cos(v_s t_2/R_s)] \tag{13}$$

According to the boundary conditions:$v_s t_1/R_s = 2\pi\text{-}v_s t_2/R_s$, $\sin(v_s t_1/R_s) \approx v_s t_1/R_s$, solve Eqs. (12) and (13), we can obtain

$$t_1 = \pi\frac{v_w}{v_s}\frac{R_s}{v_s + v_w} \tag{14}$$

$$t_2 = \frac{\pi R_s}{v_s}\left(2\text{-}\frac{v_w}{v_s + v_w}\right) \tag{15}$$

The value of $t_1$ and $t_2$ is compared with each others to identify which is the moment grit reaches point $B$ the first time, and then

$$t_2\text{-}t_1 = \frac{2\pi R_s}{v_s}\left(1\text{-}\frac{v_w}{v_s + v_w}\right) > 0 \tag{16}$$

Therefore, $t_1$ is the first moment grit reaches point $B$ and $t_2$ is the second moment.

### 2.1.4 Contact arc length between the grain and workpiece

The equation of the unit contact arc length *dl* during the grinding process is

$$dl = \sqrt{dx^2 + dy^2} = \sqrt{v_s^2 + 2v_s v_w \cos(v_s t/R_s) + v_w^2}dt \tag{17}$$

So, this is simplified as

$$dl \approx (v_s + v_w)dt \tag{18}$$

Therefore, the contact arc length for one cutting process is

$$l = \int_{t_2}^{t_{1\max}} dl = \int_{t_2}^{t_{1\max}} (v_s + v_w)dt = \frac{R_s(v_s + v_w)}{v_s}\left[\arccos\left(1 - \frac{a_p}{R_s}\right) + \frac{\pi v_w}{v_s + v_w}\right] \tag{19}$$

Based on the angle relation between $\alpha$ and $a_p$, $R_s$.M eanwhile,$a_p \ll R_s$, so $a_p{}^2/R_s{}^2 \approx 0$, $\sin\alpha \approx \alpha$. Then:

$$\alpha \approx \sqrt{2a_p/R_s - a_p{}^2/R_s{}^2} \approx \sqrt{2a_p/R_s} \tag{20}$$

Then:

$$l = \frac{R_s(v_s + v_w)}{v_s}\left(\sqrt{\frac{2a_p}{R_s}} + \frac{\pi v_w}{v_s + v_w}\right) = \sqrt{2a_p R_s} + \frac{v_w}{v_s}\left(\sqrt{2a_p R_s} + \pi R_s\right) \tag{21}$$

### *2.1.5 Removal volume for one cut*

Based on **Figure 1(a)**, the removal volume for one grit is

$$V = \int_{t_2}^{t_{1\max}} s(t)dl - \int_{t_1}^{t_{0\max}} s(t)dl = \int_{t_2 + t_{0\max}}^{t_1 + t_{1\max}} s(t)dl \tag{22}$$

Substituting Eqs. (8), (11), (14), (15) and (18) into Eq. (22), then

$$V = \int_{t_2 + t_{0\max}}^{t_1 + t_{1\max}} a_p{}^2 \tan\theta (v_s + v_w)dt = 2\pi R_s \tan\theta a_p{}^2 v_w / v_s \tag{23}$$

### *2.1.6 Specific grinding energy model*

During one cutting process, the total work single grit does is $E = F_t(v_s + v_w)(t_{1\max} - t_2)$,where $F_t$ is the tangential grinding force, and the total removal volume is $V$ as Eq. (23) shows. Then, the specific grinding energy is

$$u' = E/V = \frac{F_t(v_s + v_w)}{2a_p{}^2 v_w \tan\theta} \cdot \frac{\arccos(1 - a_p/R_s) + \pi v_w/(v_s + v_w)}{\pi} \tag{24}$$

The general specific grinding energy $u = \frac{F_t(v_s + v_w)}{v_w a_p b}$ in which b is the width of grit cutting into the workpiece when grinding depth is ap for single-grain grinding. $b = 2a_p \tan\theta$, so:

$$u = \frac{F_t(v_s + v_w)}{2a_p{}^2 v_w \tan\theta} \tag{25}$$

### 2.1.7 Grinding force model of single abrasive

There are two fiber orientations in both grinding directions, respectively: perpendicular to the fiber bundle end and side surface in direction 1, and perpendicular and parallel to the fiber bundle side surface on direction 2. **Figure 2** shows the relation between the fiber orientation of WCMC and the grinding direction.

Here are some instructions: the required energies removing unit volume perpendicular to the fiber bundle end and side surface are $u_{end}$ and $u_{per1}$ on direction 1, respectively. The removing unit volumes parallel and perpendicular to the fiber bundle side surface are $u_{per2}$ and $u_{par}$, and the volume ratios of the fiber bundle end surface on direction 1 and parallel to the fiber bundle side surface on direction 2 are β and γ separately, then

On direction 1:

$$V_{end} = \beta V, V_{per1} = (1-\beta)V \tag{26}$$

On direction 2:

$$V_{par} = \gamma V, V_{per2} = (1-\gamma)V \tag{27}$$

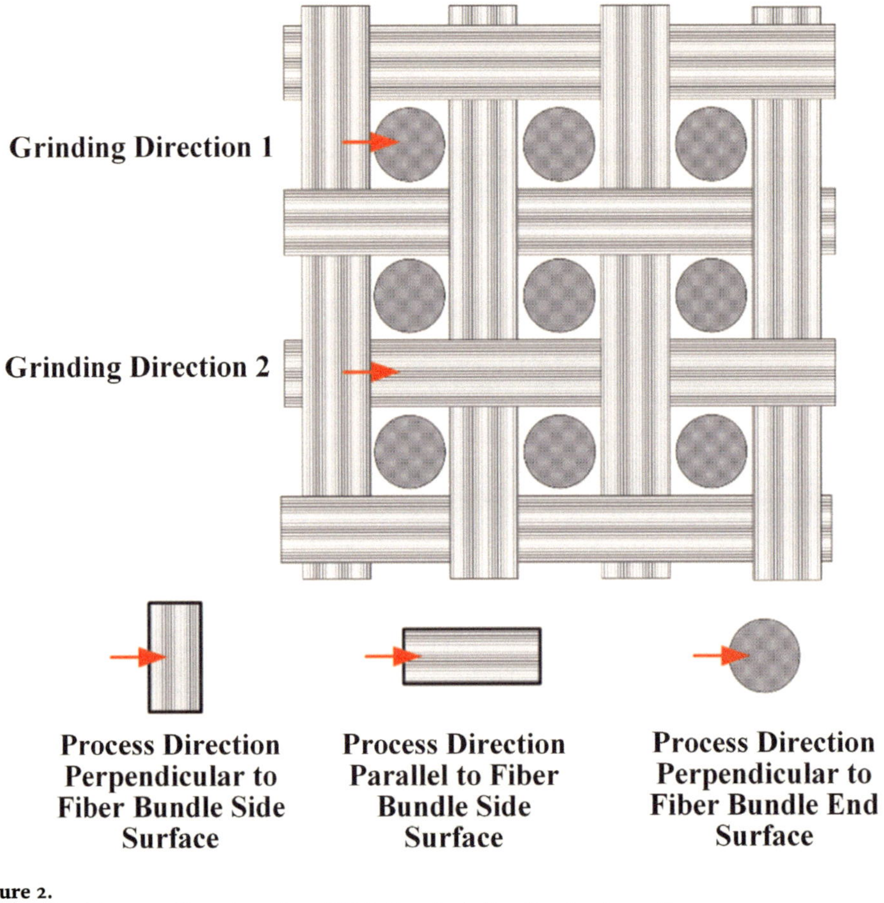


**Figure 2.**
*Relation sketch between fiber orientation of WCMC and grinding direction [25, 26].*

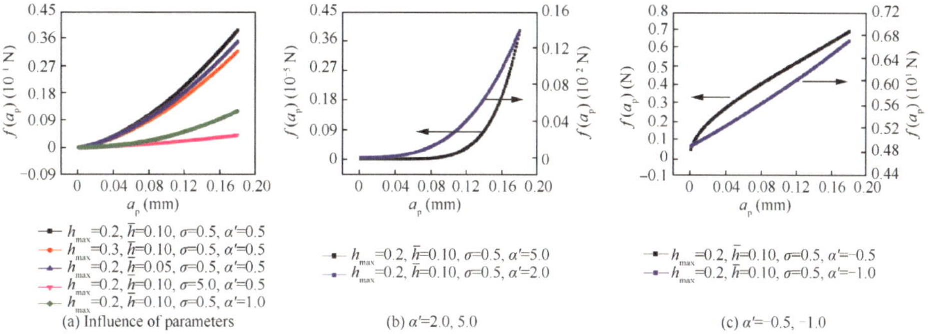


**Figure 3.**
*Trends of numerical solution of tangential force with grinding depth [6].*

where $V_{end}$ and $V_{per1}$ are the removal volumes of perpendicular to the fiber bundle end and side surface on direction 1, and $V_{par}$ and $V_{per2}$ are the removal volumes of parallel and perpendicular to the fiber bundle side surface on direction 2.

So, on directions 1 and 2, the needed energy removing volume $V$ is

$$E_1 = V_{end}u_{end} + V_{per1}u_{per1} = u_{end}\beta V + u_{per1}(1-\beta)V \tag{28}$$

$$E_2 = V_{par}u_{par} + V_{per2}u_{per2} = u_{par}\gamma V + u_{per2}(1-\gamma)V \tag{29}$$

According to energy balance $E = F_t l$, substituting Eqs. (21), (23), (28) and (29), then.

On direction 1:

$$F_t = E_1/l = \frac{2\pi R_s \tan\theta a_p{}^2 v_w \left[u_{end}\beta + u_{per1}(1-\beta)\right]}{\sqrt{2a_p R_s}(v_w + v_s) + \pi R_s v_w} \tag{30}$$

On direction 2:

$$F_t = E_2/l = \frac{2\pi R_s \tan\theta a_p{}^2 v_w \left[u_{par}\gamma + u_{per2}(1-\gamma)\right]}{\sqrt{2a_p R_s}(v_w + v_s) + \pi R_s v_w} \tag{31}$$

In the formulas above, $u_{end}$, $u_{per1}$, $u_{per2}$, and $u_{par}$ are all related to processing parameters and material properties of each fiber orientation. If the composite is not 3D orthogonal, fiber orientation cannot be identified obviously, assuming that $u_f = u_{end}\beta + u_{per1}(1-\beta)$ or $u_f = u_{par}\gamma + u_{per2}(1-\gamma)$, then:

$$F_t = \frac{2\pi R_s \tan\theta a_p{}^2 v_w u_f}{\sqrt{2a_p R_s}(v_w + v_s) + \pi R_s v_w} \tag{32}$$

where $u_f$ is the specific energy related to grinding direction and fiber orientation. Then, the normal force $F_n$ can be obtained through Eq. (33) as follows:

$$F_n = (F_t \pm C_F)/\mu \tag{33}$$

Here, $\mu$ is the friction coefficient among grinding grain, processed material surface, and debris, and $C_F$ is a constant that is related to the grinding process.

## 2.2 The analysis and modeling of grinding force

Based on Eqs. (32) and (33), the theoretical model of tangential force during grinding is obtained by introducing the randomness of abrasive particles [6]:

$$\overline{F_t} = E\left(\frac{\xi \tan\theta' {a_p}^{\alpha} {v_s}^{\beta} {v_w}^{\gamma}}{\sqrt{2a_p R_s}(v_w + v_s) + \pi R_s v_w}\right) = -\overline{\xi}(\overline{\tan\theta'})E\left(\frac{{a_p}^{\alpha} {v_s}^{\beta} {v_w}^{\gamma}}{\sqrt{2a_p R_s}(v_w + v_s) + \pi R_s v_w}\right) \quad (34)$$

where $E(\cdot)$ refers to a mathematical expectation; $\overline{\xi}$ and $\overline{\tan\theta'}$ are the expectations of $\xi$ and $\tan\theta$, respectively. $R_s$ is the grinding wheel radius; $\xi$ is a constant related to $R_s$; $\alpha$, $\beta$, and $\gamma$ are the fitting coefficients of $a_p$, $v_s$, and $v_w$ with actual processing, which is related to grinding direction and fiber orientation; $\theta'$ is half of the single-grain tip angle. Therefore, once the grinding wheel and the workpiece are selected, the width b, the radius, the grains exposure height distribution, and the grains protrusion angle are invariable. For this reason, the working conditions and the grinding wheel state are practically constant, given that the wheel wear is ignored. Thus, the expected $\xi$ and $\theta'$ are constant.

Since $v_w \ll v_s$, $a_p \ll R_s$ for each grain, meaning that the minuscule changes of the true Rs and vs. are eliminated. When the nominal $v_s$ and $v_w$ values are constant during processing and ap is variable, Eq. (34) can be simplified as

$$\overline{F_t} = \xi' \tan\overline{\theta'} E\left({a_p}^{\alpha'}\right) \quad (35)$$

where $\xi'$ is a constant describing the grinding wheel condition and the machining process and processing parameters $v_s$ and $v_w$, while $\alpha'$ are the fitting coefficient of $a_p$.

$$a_{pi} = h_i - (h_{\max} - a_p) \quad (36)$$

According to the law of large numbers, the grain exposure height $h_i$ distribution follows a normal distribution:

$$h_i \sim N\left(\overline{h}, \sigma^2\right) \quad (37)$$

where $\overline{h}$is the average value of $h_i$, $\sigma$ is the standard error, and the actual depth $a_{pi}$ also follows a normal distribution:

$$a_{pi} \sim N\left(\overline{h} - h_{\max} + a_p, \sigma^2\right) \quad (38)$$

Eq. (38), it can be divided into two cases.

The first is $h_{\max} > a_p$, under which only $h_i \geq h_{\max} - a_p$, the grains can participate in the grinding. Therefore, Eq. (34) can be written as

$$\overline{F_t} = \xi' \tan\overline{\theta'} \int_0^{a_p} {a_{pi}}^{\alpha'} \frac{1}{\sqrt{2\pi}\sigma} e^{-\frac{\left[a_{pi} - \left(\overline{h} - h_{\max} + a_p\right)\right]}{2\sigma^2}} da_{pi} \quad (39)$$

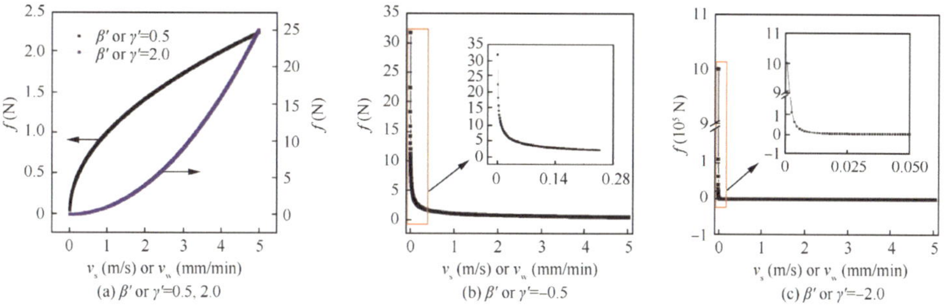


**Figure 4.**
*Trends of numerical solution of tangential force with wheel speed or feed rate [6].*

Eq. (39) is a transcendental equation with too many variables to obtain an analytical solution. So, assuming:

$$f\left(a_p\right) = \int_0^{a_p} a_{pi}{}^{\alpha'} \frac{1}{\sqrt{2\pi}\sigma} e^{-\frac{\left[a_{pi}-\left(\overline{h}-h_{\max}+a_p\right)\right]}{2\sigma^2}} da_{pi} \tag{40}$$

The trends of the numerical solution of tangential force with grinding depth are shown in **Figure 3**, reflecting that independently of the changes in other parameters, the value of $f(a_p)$ increases with $a_p$. Thus, $\overline{F}_t$ also increases.

When the nominal value of $a_p$ is constant, $h_{\max} > a_p$ and $h_i \geq h_{\max} - a_p$ during processing. $v_s$ or $v_w$ is changing, meaning that Eq. (34) can be simplified as

$$\overline{F_t} = \xi' \tan \overline{\theta'} v_s{}^{\beta'} \tag{41}$$

$$\overline{F_t} = \xi' \tan \overline{\theta'} v_w{}^{\gamma'} \tag{42}$$

where $\beta'$ and $\gamma'$ are the fitting coefficients $v_s$ and $v_w$. Assuming that $f = v_s{}^{\beta'}, f = v_w{}^{\gamma'}$, the numerical solution variation trend of the variable $f$ is as shown in **Figure 4**.

On the other hand, when $h_{\max} < a_p$, all the grains and the wheel matrix take part in the grinding process. Thus, the tangential force will have two components: the first caused by the interaction between the grain and workpiece, and the second caused by the interaction between the wheel matrix and workpiece. The resulting situations for the tangential and normal forces are more complicated than those of single-grain grinding.

## 3. Tests and methods

The grinding type used in this study is up-grinding without coolant. The dynamometer is Kistler 9257A, and the data acquisition system is Kistler 5070A which was used with the dynamometer sampling frequency of 20 kHz. Grinding experiments are performed on the JDUT400E CNC machine. The workpiece material is 3D orthogonal SiO2f/SiO2. After the experiment, the surface roughness Sa is obtained via a non-contact 3D optical measurement instrument NANOVEA ST400. In the single-particle scratching test, the grinding wheel is customized with a single diamond grain, and the radius of rotation during abrasive processing is 13.6 mm. The specimens have the dimensions of 50 × 25 × 15 mm, and grinding are along the longest direction with the grain engaged across the entire workpiece length. The real grinding process is shown in **Figure 5(a)**. In the grinding test, the tool is specially designed as an electroplated diamond with grain numbers of 100 # and 240 #. The wheel diameter is 16 mm, which was used to carry out all the grinding experiments. There are two fiber orientations on the processing surface. Since the grinding forces of fibers in different directions vary, the wheel width of 7 mm was selected to ensure that the grinding zone includes two areas containing two different fiber orientations. The specimens have dimensions of 40 × 27 × 25 mm. The front surface is the grinding surface, while the grinding direction is along the 40 mm side. All the workpiece surfaces are ground to planes in advance, guaranteeing that, independently of the grinding depth, machining does not cross the 1 mm thickness fiber layer. The real grinding process is shown in **Figure 5(b)**.

The process parameters of the single particle scratching test and grinding wheel test are shown in **Table 1** and **Table 2**, respectively. In the single-particle scratch test, the grinding experimental parameter design adopts single-factor grinding. In the grinding test, a full factorial experiment is adopted. The workpiece feed rate has a lower influence on the processing force than the grinding speed and depth. Thus, the workpiece feed rate is set to 500 mm/min. In order to compare with the grinding experiment of single grit, the selected grinding wheel speed and grinding depth level are the same as that of single grit. And each experimental run (for each of the conditions) is replicated three times.

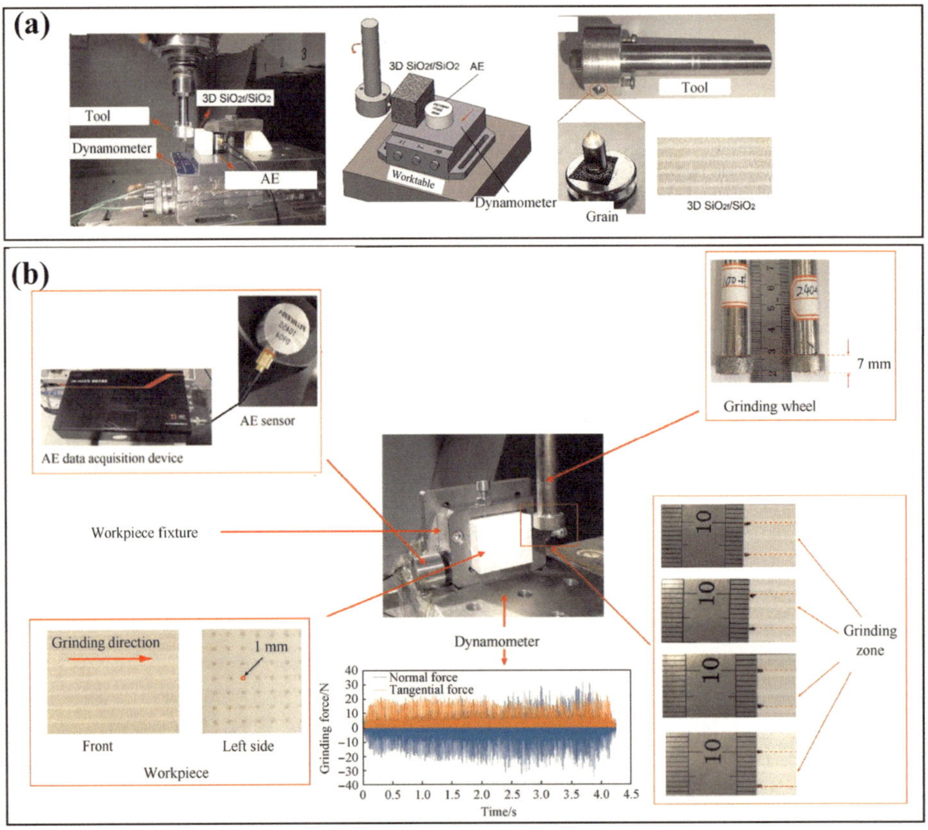


**Figure 5.**
*Experiment setup: (a) the single particle scratch test and (b) the grinding experiment setup [6, 25].*

| Parameters | Value |
|---|---|
| Grinding wheel radius *Rs* (mm) | 13.6 |
| Grain vertex half angle $\theta$ (°) | 38.7 |
| Grinding wheel speed $v_s$ (r/min) | 1000, 1500, 2000, 2500, 3000 |
| Feed rate of workpiece $v_w$ (mm/min) | 500, 550, 600, 650, 700 |
| Grinding depth $a_p$ (mm) | 0. 1, 0. 12, 0. 14, 0.16, 0. 18 |

**Table 1.**
*Grinding parameters of the single particle scratch test.*

| Parameters | | | Value | | |
|---|---|---|---|---|---|
| Feed rate (mm/min) | | | 500 | | |
| Grinding wheel | 100# | | | 240# | |
| Grinding speed (m/s) | 1.4 | 2.1 | 2.8 | 3.5 | 4.2 |
| Grinding depth (mm) | 0.1 | 0.12 | 0.14 | 0.16 | 0.18 |

**Table 2.**
*Grinding parameters of the full factor experiment [6].*

# 4. Result and discussion

## 4.1 Force analysis on a single abrasive particle

### *4.1.1 Specific grinding energy*

Through Eqs. (24) and (25), the specific grinding energies of our proposed model and the general model are obtained, as shown in **Figure 6**. Experimental result demonstrates that the specific grinding energy sharply decreases with increasing wheel speed and increases slightly with the increasing feed rate and grinding depth in both processing directions. The specific energy of direction 1 is larger than that of direction 2 under the same processing parameters. The proposed specific energies and the general ones have the same changing trends as each other under grinding parameters, but the values and the order of magnitude have a huge difference.

The difference between the general and the proposed models are reflected in two aspects. One is the model in the research and is based on single grain which is always taking part in the machining of each cut. The grain edge height is constant relative to the workpiece when the grinding depth is fixed. The other is the model that considers the contact period from $t_2$ to $t_{1\ \max}$ between the grain and workpiece in one cut. It can be known that from the moment $t_{1max}$ to the next time the grain cutting into the workpiece, single grain does not take part in grinding anymore. Therefore, the angle ratio comparing Eqs. (24) and (25) appears, and the specific energy is smaller. Furthermore, each grain taking part in grinding does not always have contact with the workpiece during conventional wheel grinding, but when it cuts out of the workpiece or is within the cutting process, other grains are participating in cutting. Therefore,

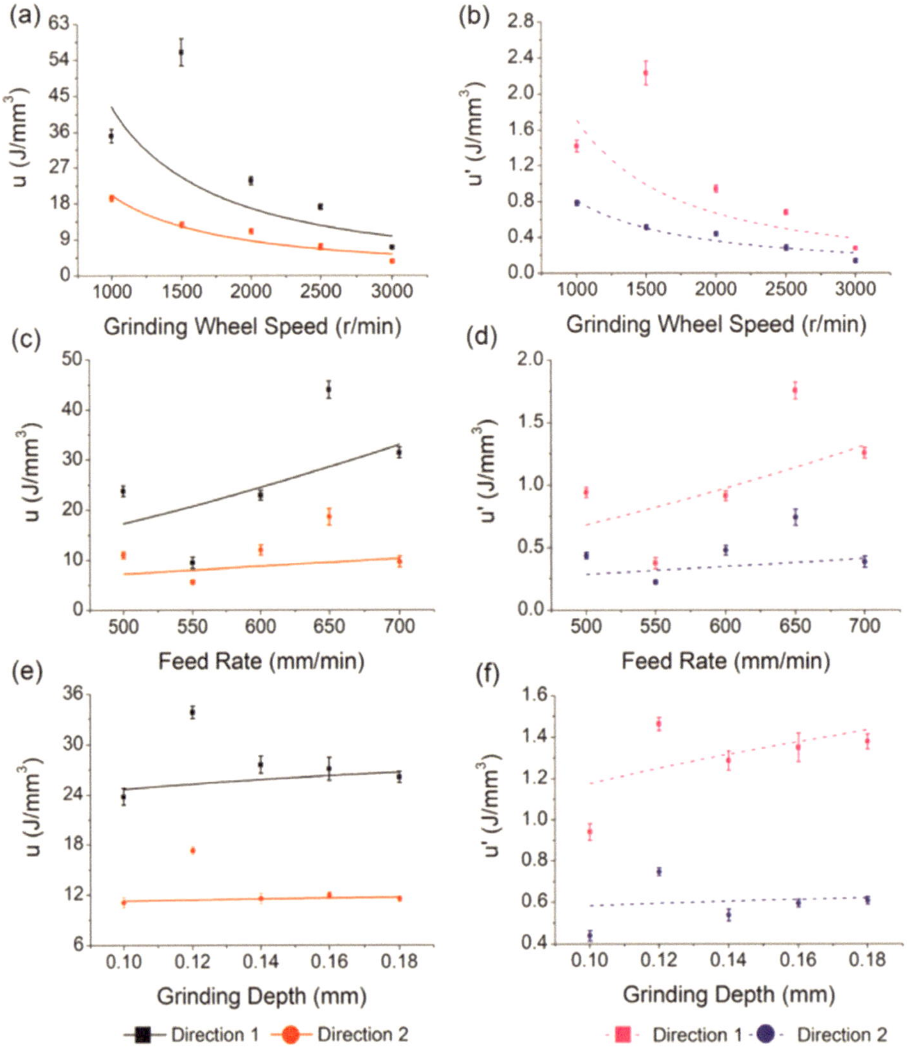


**Figure 6.**
*Trends of specific grinding energy with different grinding parameters: (a), (c), (e) general specific energy. (b), (d), (f) proposed specific energy [25].*

for conventional wheel grinding, the process is successive on a macroscopic scale; but for single grinding, the process is discontinuous no matter on a macroscopic or microscopic scale.

### *4.1.2 Force model verification*

**Figure 7** is the experimental and fitting results of the force with changing grinding parameters, which show good accordance with the experimental ones, and reflects the rationality of the force model. Meanwhile, the grinding force has a similar trend with the change of specific energy when the machining parameters change. This is because force is closely related to changes in energy. The grinding process is steadier, and the

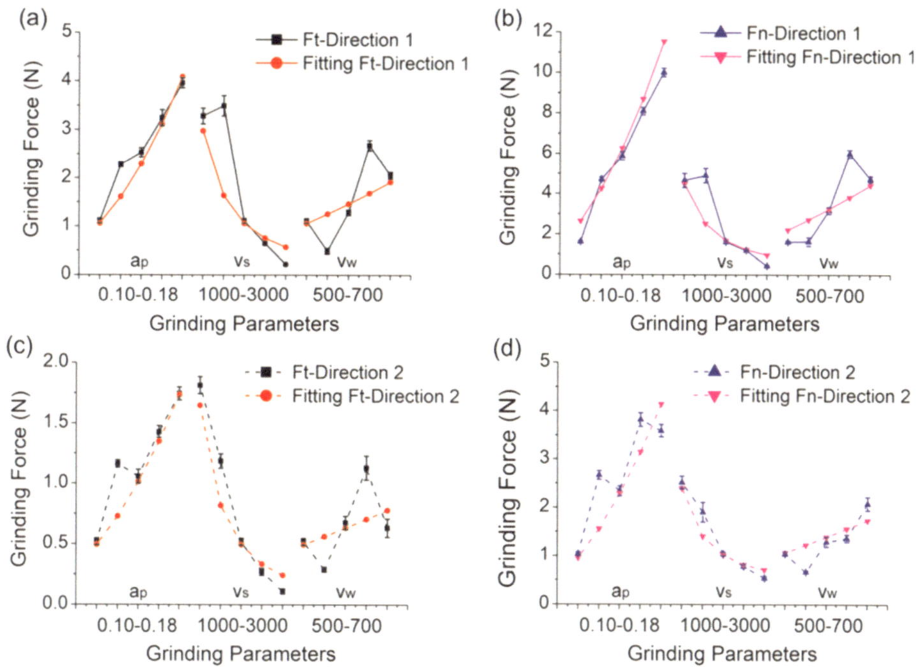


**Figure 7.**
*Experimental and fitting results: (a)* $F_t$ *on direction 1, (b)* $F_n$ *on direction 1, (c)* $F_t$ *on direction 2, and (d)* $F_n$ *on direction 2 [25].*

specific energy and the grinding force decrease. Inversely, increasing the feed rate will reduce the brittle fracture and increases the plastic deformation of the material resulting in a larger specific energy and grinding force. Increasing cutting depth will increase the material removal rate. Therefore, the integrated effects make the specific energy and the grinding force increase. From the global, the force and the specific energy are not just going up or down, and there are some fluctuations with the grinding parameters. The main reason is that the presence of reinforced fibers causes their fluctuations [27].

## 4.2 Grinding force analysis of grinding wheel

### *4.2.1 Effect of process parameters on grinding force*

The grinding force contour maps with the change of grinding speed and depth are shown in **Figure 8**. It can be seen that the changes of tangential and normal forces are similar for the 100 # grinding wheel. A peak and closed profile was found in **Figure 8 (a)** and **(b)**, meaning that there is a grinding force turning point. Furthermore, the difference is revealed between the maximum values measured for tangential and normal forces. The largest tangential force was measured for a combination of the grinding speed 1.9–2.9 m/s and grinding depth 0.145–0.168 mm. On the other hand, the largest normal force was found for a combination of the grinding speed 2.7–3.0 m/s and grinding depth 0.117–0.139 mm. Generally speaking, the grinding force increases as the grinding speed decreases and the grinding depth increases, but there are some differences here. This phenomenon could be attributed to the change of fiber

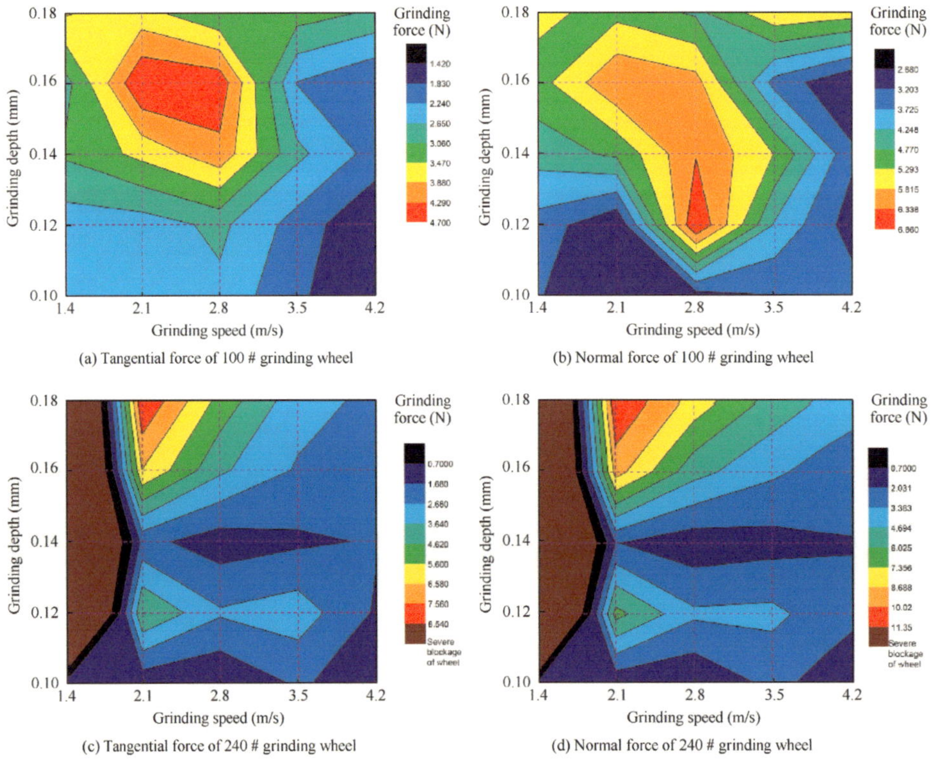


**Figure 8.**
*Grinding force contour maps with the change of grinding speed and depth [6].*

cutting types and tribological behavior between the abrasive grains and $SiO_2f/SiO_2$. With the increase in grinding depth and decrease in grinding speed, the wear debris is crushed into small pieces, which changes the $SiO_2$ fiber cutting type, and causes a decrease in grinding force. The contour maps of grinding force using a 240 # grinding wheel are in **Figure 8(c)** and **(d)**. Both tangential and normal forces are high when the grinding speed ranges from 2.0 to 2.4 m/s, and the grinding depth is 0.165–0.180 mm. Furthermore, when the grinding speed ranges from 1.4 to 1.9 m/s, the grinding depth is above 0.100 mm, the grinding wheel encounters a severe blockage.

### *4.2.2 Comparison of the force of the single particle and grinding wheel*

The grinding force for both the single-grain and grinding wheel grinding using the same parameters were shown in **Figure 9**. When the grinding depth is changed from 0.10 to 0.18 mm, the grinding speed is set to 2.8 m/s. The grinding speed changes from 1.4 to 4.2 m/s, while the grinding depth is set to 0.10 mm. The experimental results indicate that both the tangential and normal forces in single-grain grinding decrease with the grinding speed and increase with the grinding depth, as discussed in the previous study by the authors. However, the diamond wheel grinding performance is notably different. The diamond wheel grinding force fluctuates with the increases in grinding speed and grinding depth, particularly for 240 # grinding wheel. When testing the grinding force using a grinding wheel, the grains on the grinding wheel are much more complicated than the single-grain test. For example, the heights and conditions of the gains on the grinding wheel are different from each other. Thus, the contacting

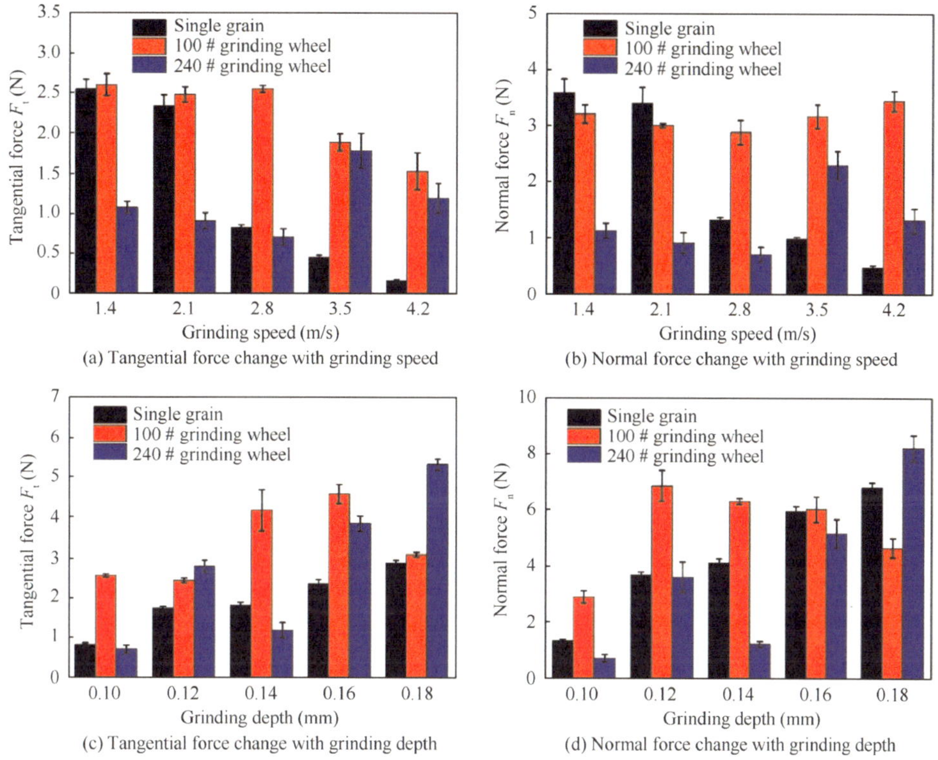


**Figure 9.**
*Force of single-grain and diamond wheel grinding [6].*

condition between the grinding wheel and the material surface is much more complicated than single-grain grinding. Due to the complexity of the grinding wheel diamond grains, the material removal type of grinding wheel will be significantly more complicated than single-grain scratching, that is the reason why the grinding performance of single-grain grinding and grinding wheel grinding is different.

### *4.2.3 Surface roughness after grinding*

It was found that the surface roughness following the grinding using 240 # grinding wheel is notably smaller compared to 100 # grinding wheel, as shown in **Figure 10**. The difference in grain size of the two grinding wheels causes the difference in surface roughness. Larger grains cause wider and deeper scratches on the workpiece, leading to a rougher grinding surface. The grain size of 100 # grinding wheel is 124–178 um, and that of 240 # grinding wheel is 53–74 um.

Furthermore, with an increase in grinding speed, the surface roughness obtained using 100 # grinding wheel first increased and then decreased. The surface roughness in the 240 # grinding wheel first decreased and then increased. The changes trend of surface roughness and the grinding force was the same, indicating that a smoother surface is obtained when using a smaller grinding force. It should be noted that for 240 # grinding wheel under the grinding speed of 1.4 m/s, only the test with 0.1 mm grinding depth could be performed. The remaining four tests have failed due to severe grinding wheel blockage. Thus, no surface roughness data were shown in **Figure 10(b)**.

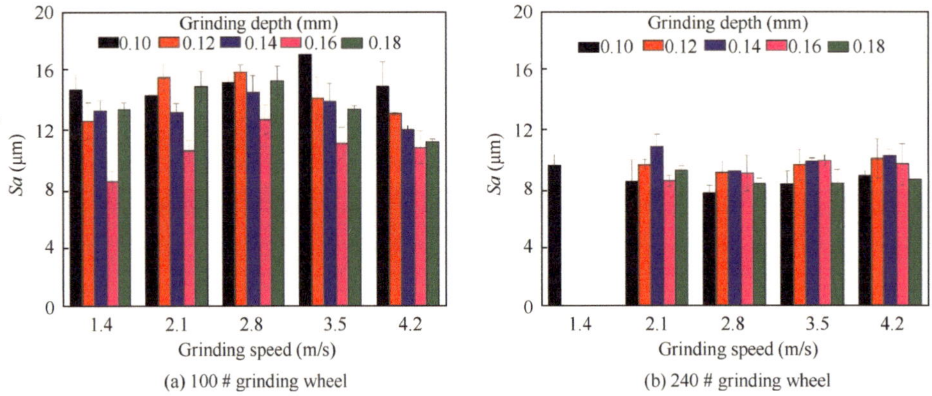


**Figure 10.**
*Surface roughness* Sa *under the different grinding speeds and depths [6].*

| Wheel | Parameters | | | | Optimum grinding parameters | | | | | |
|---|---|---|---|---|---|---|---|---|---|---|
| | $F_{min}$ (N) | | $Sa_{min}$ (µm) | | | | | | | |
| | $v_s$ (m/s) | $a_p$ (mm) | $v_s$ (m/s) | $a_p$ (mm) | $v_s$ (m/s) | $a_p$ (mm) | $Z_w$ ($mm^3/s$) | $F_t$ (N) | $F_n$ (N) | $Sa$ (µm) |
| 100# | ≥ 3.5 | ≤ 0.16 | — | 0.14–0.175 | ≥ 3.5 | 0.14–0.16 | 8.17–9.33 | ≤ 3 | ≤ 5.3 | ≤ 15 |
| 240# | ≥ 2.8 | ≤ 0.16 | 1.75–4.2 | ≥0.16 or ≤ 0.12 | 2.8–4.2 | ≤ 0.12 | ≤7 | ≤ 3.6 | ≤ 4.7 | ≤ 9.6 |

**Table 3.**
*Optimum grinding parameters of both wheels.*

### *4.2.4 Optimum grinding parameter*

**Table 3** shows the processing parameters corresponding to the minimum grinding force and surface roughness [28–30] of the two grinding wheels, respectively. By taking the intersection, the optimal machining parameters of the two grinding wheels can be obtained. At the same time, the respective nominal removal rates are calculated under the optimal processing parameters. Reasonable machining parameters should not only ensure the machining quality and reduce the damage to the workpiece and the tool, but also consider the processing efficiency.

## 5. Conclusions

In this chapter, a new force model of WCMC considering fiber orientation was proposed based on energy balancing theory and verified with the grinding experiment of 3D orthogonal $SiO_2f/SiO_2$. The main findings are listed as follows:

1. The study modifies the specific grinding energy model and proposes an analytical model to predict WCMC grinding force in single-grain processing in terms of fiber orientation.

2. Through the grinding experiments, the grinding force and the specific energy decrease with the increasing wheel speed but increase with the increasing feed speed and grinding depth.

3. During the grinding process, the force and specific energy of direction 1 are always larger than those of direction 2, and the normal force is always bigger than the tangential force.

4. Aiming at minimum grinding force and maximum surface quality, the optimum process parameters of two grinding wheels are obtained.

The research content of this paper includes some published papers, which have attracted wide attention. We will continue to carry out extensive research on the processing of ceramics in the future, especially the processing of woven ceramic matrix composite and ceramic electronic components.

## Acknowledgements

This work was supported by the National Natural Science Foundation of China [51375333].

## Conflict of interest

The authors declare no conflict of interest.

## Author details

Bin Lin[1,2]*, Jingguo Zhou[1,2], Jinhua Wei[3], Haoji Wang[1,2] and Tianyi Sui[1,2]*

1 School of Mechanical Engineering, Tianjin University, Tianjin, China

2 Key Laboratory of Advanced Ceramics and Machining Technology, Ministry of Education, Tianjin University, Tianjin, China

3 Aerospace Research Institute of Materials and Processing Technology, Beijing, China

*Address all correspondence to: tdlinbin@126.com; suity@tju.edu.cn

## References

[1] Chen D, Liu Y, Meng M, et al. Dynamic axial crushing behaviors of circular composite tubes with different reinforcing fibers and triggers. International Journal of Mechanical Sciences. 2023;**244**:108083. DOI: 10.1016/j.ijmecsci.2022.108083

[2] Wang H, Koyanagi T, Arregui-Mena DJ, et al. Anisotropic thermal diffusivity and conductivity in SiC/SiC tubes studied by infrared imaging and X-ray computed tomography. Ceramics International. 2022;**48**(15):21717-21727. DOI: 10.1016/j.ceramint.2022.04.153

[3] Pan Z, Qiao F, Yu J, et al. Distribution of axial yarns on the localized deformation and damage mechanism of triaxial braided composite tubes. Thin-Walled Structures. 2022;**177**:109389. DOI: 10.1016/j.tws.2022.109389

[4] Park JY, Kim D, Lee HG, et al. Fabrication and evaluation of C-ring strength of SiCf/SiC composite tube. Journal of the Korean Ceramic Society. 2021;**58**(6):718-727. DOI: 10.1007/s43207-021-00146-0

[5] Teti R. Machining of composite materials. CIRP Annals - Manufacturing Technology. 2002;**51**(2):611-634. DOI: org/10.1016/S0007-8506(07)61703-X

[6] Lin B, Wang H, Wei J, et al. Diamond wheel grinding characteristics of 3D orthogonal quartz fiber reinforced silica ceramic matrix composite. Chinese Journal of Aeronautics. 2021;**34**(5):404-414. DOI: 10.1016/j.cja.2020.12.026

[7] Yao L, Liu Z, Song Q, et al. Prediction modelling of cutting force in rotary ultrasonic end grinding 2.5D woven $SiO_2f/SiO_2$ ceramic matrix composite. Composite Structures. 2023;**304**(2023):116448. DOI: 10.1016/j.compstruct.2022.116448

[8] Yang Z, Zhu L, Lin B, et al. The grinding force modeling and experimental study of ZrO2 ceramic materials in ultrasonic vibration assisted grinding. Ceramics International. 2019;**45**(7):8873-8889. DOI: 10.1016/j.ceramint.2019.01.216

[9] Li C, Li X, Wu Y, et al. Deformation mechanism and force modelling of the grinding of YAG single crystals. International Journal of Machine Tools and Manufacture. 2019;**143**:23-37. DOI: 10.1016/j.ijmachtools.2019.05.003

[10] Li H, Yu T, Wang Z, et al. Detailed modeling of cutting forces in grinding process considering variable stages of grain-workpiece micro interactions. International Journal of Mechanical Sciences. 2016;**126**:319-339. DOI: 10.1016/j.ijmecsci.2016.11.016

[11] Meng Q, Guo B, Zhao Q, et al. Modelling of grinding mechanics: A review. Chinese Journal of Aeronautics. 2022. DOI: 10.1016/j.cja.2022.10.006

[12] Zhang Y, Lin B, Liu J, et al. An experimental study on mechanical modeling of ceramics based on microstructure. Applied Sciences. 2015;**5**(4):1337-1349. DOI: 10.3390/app5041337

[13] Wang Y, Sarin VK, Lin B, et al. Feasibility study of the ultrasonic vibration filing of carbon fiber reinforced silicon carbide composites. International Journal of Machine Tools and Manufacture. 2016;**101**:10-17. DOI: 10.1016/j.ijmachtools.2015.11.003

[14] Huang H, Liu YC. Experimental investigations of machining characteristics and removal mechanisms

of advanced ceramics in high speed deep grinding. International Journal of Machine Tools & Manufacture. 2003;**43**: 811-823. DOI: 10.1016/S0890-6955(03) 00050-6

[15] Zhang Y, Li C, Ji H, et al. Analysis of grinding mechanics and improved predictive force model based on material-removal and plastic-stacking mechanisms. International Journal of Machine Tools and Manufacture. 2017;**122**:81-97. DOI: 10.1016/j.ijmachtools.2017.06.002

[16] Patnaik Durgumahanti US, Singh V, Venkateswara RP. A new model for grinding force prediction and analysis. International Journal of Machine Tools and Manufacture. 2010;**50**(3):231-240. DOI: 10.1016/j.ijmachtools.2009.12.004

[17] Li Z, Zhang F, Luo X, et al. A new grinding force model for micro grinding RB-SiC ceramic with grinding wheel topography as an input. Micromachines. 2018;**9**(8):368. DOI: 10.3390/mi9080368

[18] Yuan S, Fan H, Amin M, et al. A cutting force prediction dynamic model for side milling of ceramic matrix composites C/SiC based on rotary ultrasonic machining. The International Journal of Advanced Manufacturing Technology. 2016;**86**(1–4):37-48. DOI: 10.1007/s00170-015-8099-6

[19] Wang Y, Lin B, Zhang X. Research on the system matching model in ultrasonic vibration-assisted grinding. The International Journal of Advanced Manufacturing Technology. 2014;**70**(1): 449-458. DOI: 10.1007/s00170-013-5269-2

[20] Ding K, Fu Y, Su H, et al. Study on surface/subsurface breakage in ultrasonic assisted grinding of C/SiC composites. International Journal of Advanced Manufacturing Technology. 2017;**91**(9–12):3095-3105. DOI: 10.1007/s00170-017-0012-z

[21] Zhang J, Lin B, Fei J, et al. Modeling and experimental validation for surface error caused by axial cutting force in end-milling process. International Journal of Advanced Manufacturing Technology. 2018;**99**(1–4):327-335. DOI: 10.1007/s00170-018-2468-x

[22] Ojo SO, Ismail SO, Paggi M, et al. A new analytical critical thrust force model for delamination analysis of laminated composites during drilling operation. Composites Part B: Engineering. 2017; **124**:207-217. DOI: 10.1016/j.compositesb.2017.05.039

[23] Cao X, Lin B, Zhang X. Investigations on grinding process of woven ceramic matrix composite based on reinforced fiber orientations. Composites Part B Engineering. 2015;**71** (mar.):184-192. DOI: org/10.1016/j.compositesb.2014.11.029

[24] Cao X, Lin B, Wang Y, et al. Influence of diamond wheel grinding process on surface micro-topography and properties of $SiO_2/SiO_2$ composite. Applied Surface Science. 2014;**292**: 181-189. DOI: 10.1016/j.apsusc.2013.11.109

[25] Wei J, Wang H, Lin B, et al. A force model in single grain grinding of long fiber reinforced woven composite. The International Journal of Advanced Manufacturing Technology. 2019;**100** (1–4):541-552. DOI: 10.1007/s00170-018-2719-x

[26] Wei J, Wang H, Lin B, et al. Acoustic emission signal of fiber-reinforced composite grinding: Frequency components and damage pattern recognition. International Journal of Advanced Manufacturing Technology. 2019;**103**(1–4):1391-1401. DOI: 10.1007/s00170-019-03645-x

[27] Liu C, Ding W, Yu T, et al. Materials removal mechanism in high-speed

grinding of particulate reinforced titanium matrix composites. Precision Engineering. 2018;**51**:68-77. DOI: 10.1016/j.precisioneng.2017.07.012

[28] Wei J, Wang H, Lin B, et al. Measurement and evaluation of fiber bundle surface of long fiber reinforced woven composites. Surface Topography-Metrology and Properties. 2019;**7**(1). DOI: 10.1088/2051-672X/aaf6fd

[29] Wei J, Wang H, Lin B. Measurement of cell body and the whole surfaces of long Fiber reinforced woven composites. IOP Conference Series Materials Science and Engineering. 2019;**678**:012029. DOI: 10.1088/1757-899X/678/1/012029

[30] Wei J, Lin B, Cao X, et al. Two-dimensional evaluation of 3D needled Cf/SiC composite fiber bundle surface. Applied Surface Science. 2015;**355**: 166-170. DOI: 10.1016/j.apsusc.2015.06.182

Chapter 8

# Lime Mortars Containing Ceramic Material as Pozzolan

*Leane Priscilla Bonfim Sales, Aline Figueiredo da Nóbrega, Iranilza Costa da Silva, Ana Cecília Vieira da Nóbrega, Arnaldo Manoel Pereira Carneiro, Fabiola Luana Maia Rocha and Diego de Paiva Bezerra*

## Abstract

Lime mortars have been indicated for restoration and conservation interventions in historic buildings, however, the slow hardening of these mortars does not favor their use and dissemination in construction areas. The inclusion of pozzolans improves these properties, and although the results achieved are not close to those found in conventional cementitious mortars, they are seen as compatible materials for restoration services, since they present moderate mechanical responses and chemical compatibility. This chapter aims to show the impact of different pozzolans on fresh and hardened lime mortar's properties, including mechanical, rheological, and microstructural properties. In addition, an overview of historical mortars is presented.

**Keywords:** historical mortar, lime mortar, pozzolan, metakaolin, conservation, restauration

## 1. Introduction

Among materials found in secular constructions, lime mortars stand out and were used until at least the beginning of twentieth century [1]. Studies that analyze the coatings of old buildings show that not only aerial lime was found, but also natural hydraulic lime and lime with the addition of pozzolans [2–4].

The use of lime in mortars has been rescued and is indicated in restoration interventions in historical buildings to ensure physical and chemical compatibility of new mortars, with respect to old ones [5, 6]. This is because using Portland cement in restoration and conservation works presents incompatibilities, in view of its different characteristics in relation to original mortar, such as high rigidity, low porosity and presence of alkaline hydroxides that can react forming soluble salts [7]. Such characteristics lead to pathologies such as fissures and efflorescence.

Lime mortars have some disadvantages compared to those produced with Portland cement, such as lower mechanical strength and very long hardening time [8]. Studies show that lime and pozzolan mixture implies changes in setting and hardening

 IntechOpen

mechanism and in rheological responses, as well as in improvement in performance regarding water permeability, durability and mechanical resistance [6, 9–11].

There is evidence of use of volcanic ash and ground ceramic fragments as natural pozzolans throughout history and it was associated with obtaining resistant and durable mortars [12], but in recent works, metakaolin is among the most used pozzolans in lime-based mixtures, justified according to authors by its high pozzolanic activity [5, 6, 8–10, 13–16] and Veiga et al. [17] reinforces that lime mortars added with pozzolans in general are chemically and physically compatible with old mortars, having a similar composition and ability to accommodate the movement of masonry structures.

## 2. Historical mortars

Throughout the useful life of a structure, there is inevitably a need for conservation and restoration, in view of its exposure to environmental weather and wear due to human use.

Various types of binders were used in the past, some like clays and bitumen were ready to use, and others required heating and mixing with water before application [18]. Also according to the authors, aerial binders such as gypsum and lime were initially used. Then hydraulic binders came up, resulting from the calcination of impure lime or from the calcination of pure lime with materials with silica and alumina.

Thus, until mid-nineteenth century, lime was the main binder for laying and coating mortars and its use was frequent in ancient historical civilizations [19]. Veiga [20] mentions that aerial lime mortars are part of almost all old buildings and performed structural functions, when used in masonry blocks, to protective, adhesion and decorative functions, when used in plasters, laying mortars and paintings, respectively.

Currently, studies have been developed to know its properties from a technical point of view, recover lost technology and define application, mixing and maintenance techniques [19].

The application of mixtures based on currently conventional binder, Portland cement, presents, in conservation and restoration services in historical buildings, physical and chemical incompatibilities with original materials of structure [5], since it is associated with high mechanical resistance, high modulus of elasticity, low permeability to water vapor and the presence of alkaline hydroxides that can react forming soluble salts [7].

As for the use of cement in repairs to old buildings, Veiga [21] also points out other common drawbacks, such as the difference in surface texture and the way it reflects light, the presence of soluble salts that can be transported to interior of walls and crystallize, leading to element degradation as well as excessive stiffness and limited ability to allow wall to dry. Magalhães [22] presents old wall pathologies associating them with their possible causes.

Despite current imposition of productivity, reduction of costs and deadlines and absorption of non-specialized labor in the field of civil construction, conservation and restoration services demand care to avoid future pathologies. Compatibility between masonry materials and laying and coating mortars contribute to proper functioning of wall and increase its durability. In the past, the mix of materials for this purpose was based on local availability [23].

Therefore, in most severe cases of old coatings degradation, it is necessary to replace all original material with material that is close to previous one and that is durable [17], where the original material is usually lime-based mortars [5, 6].

Grilo et al. [15] point out that some properties are specific to lime-based materials, such as moderate mechanical properties, good performance in water vapor transmission and durability.

Veiga [21] compares performance of current and old external coatings mortars in terms of permeability. While currently, the focus is on ensuring walls watertightness, either by cutting capillarity in foundations, waterproofing and watertight window frames, the ancient walls were thick and porous and admitted water entry, but also allowed expulsion of water to be easy and fast.

Many times, the lack of knowledge about old mortars constitution and techniques applied for adequate restoration interventions lead those responsible for such services to adopt replacement of entire coating, which deprives building of its character and is a solution that presents a performance and durability inferior to that of pre-existing, as reported by Ref. [21].

This concept, in addition to being related to the conservation of history, is also linked to sustainability concept, since an inadequate solution can accelerate degradation process. The author emphasizes that it is necessary to understand degradation degree and know materials that makeup coating and their properties.

Given the current context, Sec [23] points out that it is important to encourage application of mortars compatible with existing walls, using products available on market and suitable for current situation.

## 3. Lime mortars

Although in Europe, studies involving lime or lime and pozzolan mortars are vast [5–8, 14, 16, 24–26], Veiga [20] states that lime-based mortars use is still rare in rehabilitation works and conservation of old buildings.

Veiga [20] attributes this to the need for scientific knowledge regarding local conditions and raw materials nature, lack of training on part of professionals to perform service, and difficulty of controlling planning in view of time required for application and drying of multilayers and waiting for favorable temperature conditions.

Lime can be aerial or hydraulic depending on the chemical composition of its raw material, limestone. Aerial lime, like any aerial binder, hardens by chemical reaction with carbon dioxide ($CO_2$), and limestone to be burned at a maximum of 900°C must fundamentally consist of $CaCO_3$ or $CaMg(CO_3)_2$ [23, 27]. Hydraulic lime, on the other hand, hardens in contact with water and is obtained through burning at about 1200°C of marly limestones that have up to 20% clay, and thus contain clayey compounds such as $SiO_2$, $Al_2O_3$ and $Fe_2O_3$ [23].

Still from a raw material perspective, aerial lime can be calcitic when its main constituent is calcite ($CaCO_3$); or dolomite when there is a high magnesium content in limestone and its main constituent is dolomite $CaMg(CO_3)_2$. Vinagre [27] mentions that European standard EN 459-11 (2011) determines that calcitic lime has at least 70% (by mass) of calcium oxide and magnesium oxide sum, with a maximum of 5% of magnesium oxide, and at least 55% of available lime, while dolomite must present 80% as a minimum value of these oxides sum, of which more than 5% correspond to magnesium oxide.

When limestone is subjected to burning/calcining, $CO_2$ is released and the limestone becomes quicklime CaO (calcitic lime) or CaO + MgO (dolomitic lime). Lime can be marketed as such or hydrated, and depending on the slaking process, in powder or paste form [12, 27].

Hydraulic lime can be natural, when its raw material already has the aforementioned clayey compounds, or artificial, in a pre-defined mixture of calcium hydroxide, calcium silicates and calcium aluminates [23]. Thus, the mixture of lime and pozzolana can be seen as an artificial hydraulic lime.

The Portuguese standard [28] presents types and requirements of limes aimed at applications in civil construction field. This group includes aerial and hydraulic limes. Air lime hardens in carbon dioxide presence and hydraulic lime hardens in carbon dioxide and water presence, that is, there is carbonation and hydration reaction.

Lime with hydraulic properties is subdivided into three groups: natural hydraulic lime, formulated lime, and hydraulic lime [28]. The first is a consequence of burning more or less clayey or siliceous limestones, reduced to powder by slaking with or without grinding, but it should be noted that grinding is limited to 0.1%. Formulated lime is the result of mixing aerial lime or hydraulic lime with hydraulic or pozzolanic material and, hydraulic lime, is the product of mixing lime with other materials such as cement, slag, fly ash, limestone filler and others. All hydraulic limes are further classified according to their 28 days compression resistance.

Hydrated lime has the advantages of low drying shrinkage and good vapor permeability, but it also has disadvantages such as low initial strength, slow hardening (24–48 h), low water resistance and easy dissociation in humid environments, on the other hand, hydraulic lime has moderate strength, faster hardening speed (4–12 h), good water resistance and salt erosion resistance [29].

The state of the art is vast within limes field, in view of various types presented above and possibilities found throughout analysis of old constructions. Among studies found, there are works that evaluate historic buildings and identify materials present [30], while others are focused on the characterization of mixtures based on different types of lime in the fresh and hardened state; in this context, curing conditions, types of addition, proportion of materials, types of aggregates, use of additives, rheology, among others, are included as study variables.

Veiga [20] reviews aspects related to aerial lime mortars in an attempt to understand limitations of their use in restoration interventions. The author associates the mortar performance with lime type, carbonation degree, porosity, nature and size of aggregates present, among others. A survey of compressive strengths was carried out and shows that values vary between 0.3 and 1.6 MPa, depending on mortar composition, calcination level and slaking methods. Aspects regarding its application are also presented. Some of the limitations raised involve understanding short-term degradation factors, influence of local factors and management of lime time, either regarding appropriate environmental conditions for application or regarding the waiting time between layers application.

The binder: aggregate ratio in lime mortars ranges from 1:1 to 1:2.5, with 1:3 and 1:4 also being found, as stated by Veiga [20]. Still according to the author, the proportion of 1:3 in volume has been adopted as a reference. In fact, in [6, 9, 14], among the different proportions studied is 1:3. Nogueira et al. [31] also bring considerations about aggregates packaging and paste volume, differentiating thick and thin mortars, and point to proportions of 1:2.7 and 1:1.8, respectively.

About the mixing water, Seabra et al. [7] report that aerial lime paste demands a high water content, as it consists of small particles and therefore a high specific surface area. Bakolas et al. [32] and Azeredo [33] used a water/binder factor in pure lime and lime and metakaolin paste equal to 1.0.

Therefore, lime mortars have some drawbacks such as slow hardening, high shrinkage and low strength [8]. Such limitations are alleviated with the inclusion of pozzolans in their composition, providing lime mortars with improved performance in terms of water permeability, durability and mechanical resistance [9, 10].

The improved properties observed in lime and pozzolana pastes are associated with hardening reactions of these pastes that differ from lime pastes. While the former hardens by the simultaneous occurrence of carbonation and pozzolanic reaction, lime pastes react exclusively by carbonation. In the following topic, the conditions and mechanisms for the occurrence of reactions are presented, as well as the main compounds formed in these pastes.

## 4. Mechanisms of setting and hardening

The hardening mechanisms of these materials are different: while aerial lime pastes and mortars harden with air through carbonation, those using natural hydraulic lime or lime with pozzolana harden in contact with air and water, occurring simultaneously carbonation and hydration or pozzolanic reaction [34].

Lime pastes and mortars harden due to the reaction of calcium or magnesium hydroxide with carbon dioxide in the air, this reaction is called carbonation. The product formed from this reaction is calcium or magnesium carbonate, depending on the reagent. When pozzolan is added to aerial lime pastes and mortars, hardening is achieved through the pozzolanic reaction and carbonation, and the preponderance between the processes will depend on the curing conditions of the environment and lime composition.

Carbonation is controlled by carbon dioxide diffusion at the reaction site and the main factors involved in reaction are $CO_2$ concentration, moisture content and permeability [27, 35]. The mechanism of the carbonation reaction is also presented by the authors in **Figure 1**. The carbonation reaction is very slow, it occurs from outside to inside of mass and takes about 6 months to 1 year, or depending on weather conditions, even longer [27].

The pozzolanic reaction is defined by calcium hydroxide with siliceous or silico-aluminous materials reaction in water presence and is controlled by siliceous phase dissolution that occurs in an alkaline medium [36].

Shi and Day [37] present the pozzolanic reaction mechanism in lime and pozzolan pastes which can be seen in **Figure 2**. Although mortars do not have significant resistance up to 3 days at 23°C, micrographs of this age shown by authors indicate that all pozzolan particles are covered by a layer of CSH gel.

Although Shi and Day [37] only mention C-S-H and hydrated tetracalcium aluminate (C4AH13) formation, the main compounds formed in mortars and lime and pozzolan pastes resulting from the pozzolanic reaction with calcium hydroxide are hydrated calcium aluminates and silicoaluminates such as stratlingite (C2ASH8), hydrogarnet (C3AH6), monocarboaluminate (C4ACH11), as well as C-S-H and C4AH13, [5, 6, 10, 14, 16], especially when pozzolan is metakaolin.

One of the most influential aspects in these compounds' formation, in hardening process, and in mortars characteristics is their curing condition, as presented by Azeredo et al. [10]. In this work, the authors evaluated pastes lime and metakaolin behavior resulting from the calcination of kaolin residue cured at different relative

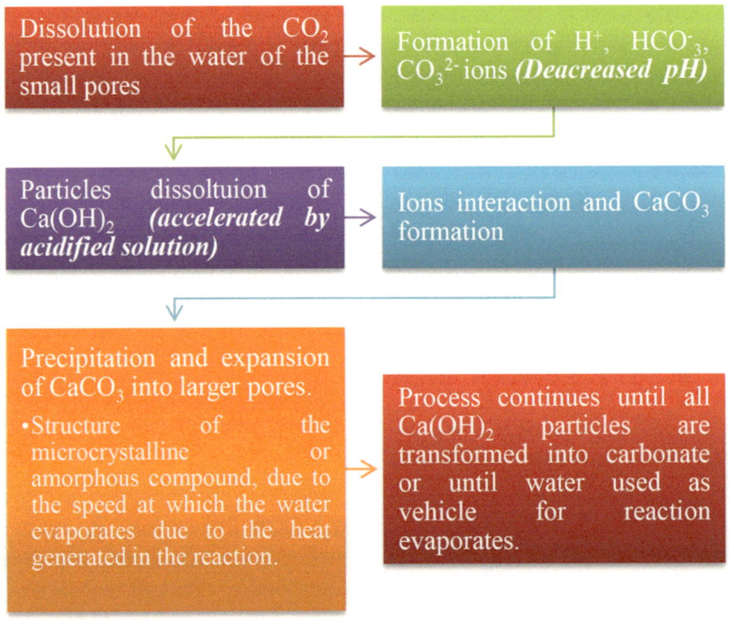


**Figure 1.**
*Mechanism of the carbonation reaction, according to Ref. [35].*

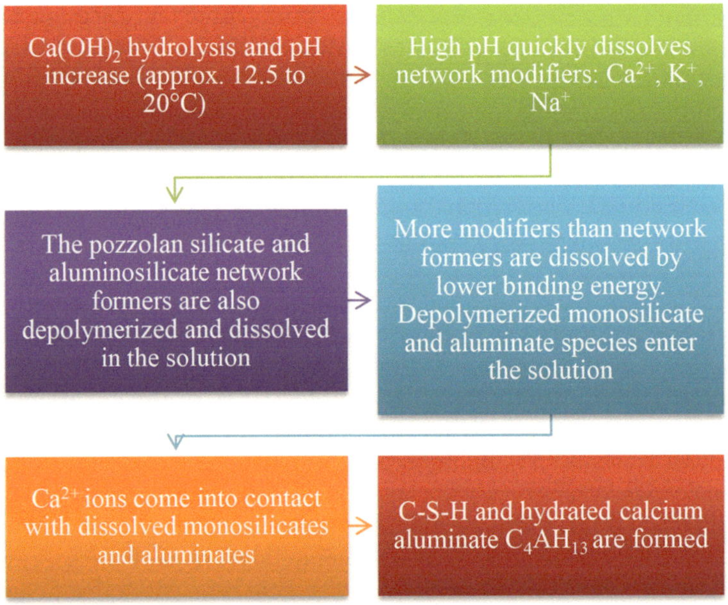


**Figure 2.**
*Mechanism of the pozzolanic reaction in lime and pozzolana pastes, according to Ref. [37].*

humidities: 65% and 100%. The results showed that wet curing favored the pozzolanic reaction to the detriment of carbonation, and was the only one that formed stratlingite. Moist curing mortars reached higher strengths.

On the other hand, as a result of substitution proportions of lime by pozzolan, the produced mortar acquires less or more hydraulic properties and influences the speed of consumption of calcium hydroxide. Gameiro et al. [5] present the replacement of 50% lime with metakaolin as the best result regarding the extension of the pozzolanic reaction up to 180 days.

The inclusion of pozzolan and the consequent setting and hardening associated with the pozzolanic reaction produces compounds with better properties, such as resistance and hardening time [6, 38]. However, it is important to emphasize that this time is still slow in relation to Portland cement-based materials [39].

## 5. Pozzolans in lime mortars

According to Ref. [23], in the past, pozzolans were classified as sand and, when they were mixed in lime mortars, gave them property of hardening under water. In the absence of natural pozzolans, from volcanic origin, Sec [23] mentions that sieved ceramic residues and "baked" clays were often used, and only in twentieth century, the use of fly ash and silica fume began.

As shown, pozzolans' inclusion in lime mortars leads to simultaneous hardening by carbonation and pozzolanic reaction in moisture presence and improves properties such as durability and setting time.

Among pozzolans, it is observed that metakaolin is pozzolanic material commonly studied in lime and pozzolan mortars and pastes. It stands out for its pozzolanic activity and consequent significant improvement in properties of these mortars [5, 6, 8–10, 13–16]. This pozzolan is derived from the calcination of kaolinite clay, between 600°C and 850°C [23].

However, there are studies that analyze mortars with lime and other ceramic materials as pozzolans, such as brick residue [40], expanded clay, rice husk ash, red brick dust and tile dust and yellow brick [41]. Besides, other materials are seen as blast furnace slag, fly ash [41] and nanosilica [26, 42].

### 5.1 Influence of pozzolans on lime mortars

#### *5.1.1 Microstructural aspects*

As seen, carbonation is responsible for calcium carbonate formation in lime mortars, however when there is a pozzolanic reaction, other compounds arise, resulting from this reaction and may vary according to mortar composition and curing conditions.

Azerêdo et al. [10] studied microstructural characteristics of hydrated lime pastes and calcined kaolin residue varying curing conditions and the lime: metakaolin ratio. The main phase formed in wet curing was stratlingite and in dry curing monocarboaluminate, in addition, wet curing and the highest content of metakaolin favored calcium hydroxide consumption. Wet curing also improved compressive strength of mortars [33]. Aspects related to setting and hardening mechanism of lime pastes and mortars were reviewed by Alvarezet al. [34].

The main compounds formed in lime and metakaolin mortars and pastes resulting from pozzolanic reaction with calcium hydroxide are hydrated calcium aluminates and silicoaluminates such as stratlingite (C2ASH8), hydrated tetracalcium aluminate (C4AH13), hydrogarnet (C3AH6), monocarboaluminate (C4ACH11), as well as hydrated calcium silicate (CSH) [5, 6, 10, 14, 16].

Moropoulou et al. [43] compared microstructural evolution of hydrated calcium silicate phase (C-S-H) in pozzolanic pastes. Powdered calcium hydroxide Ca(OH)2 was used as a reagent and, as a silicon source, silica fume or tetraethyl orthosilicate solution. The authors conclude that limiting factor in pozzolanic reaction rate is siliceous phase dissolution since CSH phase in pastes with silica fume was identified in less than 1 hour, but in pastes with tetraethyl orthosilicate in just 10 minutes.

*5.1.2 Mechanical behavior*

Velosa et al. [39] evaluated mortars with aerial lime, with aerial lime and three types of metakaolin and with cement. The results showed that there was an improvement in compressive and tensile strength and in modulus of elasticity in lime and metakaolin mixtures compared to pure aerial lime, but responses were still lower than cementitious mortar results. The authors also observed that, when comparing the three metakaolins used, the higher alumina content and the lower the alkali content, there were better mechanical results. The percentage of $Al_2O_3$ in metakaolins analyzed by authors ranged from 28 to 32%.

Veiga et al. [26] studied the application and performance of coating mortars with aerial lime and different materials, namely: hydraulic lime, white cement, natural pozzolan from Cabo Verde, metakaolin and silica fume, applied in recovery of old fortresses near the coast of Lisbon exposed to saline atmosphere, erosion by wave action, temperature variation and wind. From a mechanical point of view, in long term, *in situ* lime mortar panels with Cabo Verde pozzolan and metakaolin showed a behavior similar to that of lime and white cement mortar. Other characteristics such as resistance to water penetration and adhesion to substrate were satisfactory for mortars with metakaolin and Cabo Verde pozzolan, even though they were more permeable than those containing Portland cement. The silica fume, however, did not show enough mechanical resistance after 3 months and the authors report the lowest proportion used since they considered the highest reactivity. Thus, they suggest future studies with similar proportions for comparison.

In addition to metakaolin and Microsilica et al. [41] tested six other types of pozzolans: blast furnace slag, fly ash, expanded clay, rice husk ash, red brick dust, and tile and yellow brick dust. Among objectives of this work, in addition to comparing pozzolans performance in lime-pozzolana mortars, it sought to understand pozzolans properties and their effects on reactivity. Thus, it was observed that amorphicity is a preponderant factor for reactivity and improvement of mortar resistance, even more than silica content of the pozzolan. Of analyzed materials, metakaolin and blast furnace slag showed highest resistances, within same proportion of lime-pozzolana-sand (1:1:3 ratio). It should also be noted that metakaolin was included in pozzolans group with highest water demand, with a water/binder ratio (w/agg.) equal to 1.5, while slag was among those that required the lowest water content. (a/agl. = 0.5), and even so, metakaolin showed greater resistance.

Matias et al. [40] studied mortars with ceramic residues in different granulometries for replacement as a fine material and as an aggregate. The residues showed different pozzolanicity and consequently resulted in mechanical responses consistent with them, so the greater pozzolanic activity, the better mechanical resistance. Among the main conclusions, authors pointed out that even replacement of particulate matter such as sand proved to be a promising solution.

**Table 1** presents mechanical behavior results of lime and lime and pozzolan mortars with regard to compressive and flexural strength and elastic modulus. It is

| Author | Test age | Compressive strength (MPa) | Flexural Strength (MPa) | Elastic modulus (MPa) | Ratio/Composition *by mass **by volume |
|---|---|---|---|---|---|
| Velosa et al. [39] | 28 | 0.5 | 0.28 | 2000 | 1:3 ** (Air lime/sand) |
| | 28 | 0.75–2.6 | 0.3–0.7 | 2900–3300 | 1:0.5:2.5 ** (Air lime/ metakaolin/sand) |
| Veiga et al. [26] | 90 | 0.6 | 0.3 | 1850 | 1:1:6 ** (Air lime–hydraulic lime) |
| | 90 | 1.5 | 0.5 | 3920 | 1:0.5:2.5 ** (Air lime–Cabo Verde Pozzolan) |
| | 90 | 1.5 | 0.6 | 2550 | 1:0.25:2.5 ** (Air lime–silica fume) |
| | 90 | 1.3 | 0.4 | 2960 | 1:0.5:2.5 ** (Air lime–metakaolin) |
| | 90 | 0.7 | 0.2 | 2130 | 1:1:4 ** (Air lime–metakaolin) |
| Matias et al. [40] | 60 | 0.2 | 0.19 | — | 1:3 ** (Air lime/sand) |
| | 60 | 0.2-0.28 | 0.10-0.18 | — | 1:0.2:14.2 ** (Air lime / ceramic waste/sand) |
| | 60 | 0.28-0.30 | 0.14-0.20 | — | 1:0.5:16 ** (Air lime / ceramic waste/sand) |
| Azeredo et al. [38] | 28 | 0.55 | 0.2 | — | 1:3* (Air lime/sand) |
| | 28 | 6.6 | 1.6 | — | 1:1:6* (Air lime/ metakaolin/sand) |
| | 28 | 4.0 | 1.0 | — | 1:2:9* (Air lime/ metakaolin/sand) |
| Gameiro et al. [6] | 28 | 0.2 | 0.1 | — | 1:12 * (Air lime/sand) |
| | 28 | 3.0 | 1.15 | — | 1:1:12* (Air lime/ metakaolin/sand) |
| | 28 | 1.4 | 0.7 | — | 1:0.43:12* (Air lime/ metakaolin/sand) |

**Table 1.**
*Mechanical behavior of lime and lime and pozzolan mortars.*

emphasized that other data, with different composition ratios and test ages, can be found in respective works.

It can be seen in **Table 1**, the difference in mortar resistance when volume or mass ratio was used, so that resistances were significantly increased in mass ratio, but regardless of this, insertion of pozzolans in mortar composition was generally accompanied by improvement in mechanical resistance.

### *5.1.3 Rheological behavior*

With regard to the rheological characterization of lime and lime and pozzolana pastes, there are few studies. Fourmentin et al. [44] analyzed lime slurries and observed a better fit to the Herschel-Bulkley model, where fluid needs a flow stress to initiate flow and apparent viscosity decreases with increasing shear rate. Although in paper by Betioli et al. [45], cementitious pastes have better adjusted to model in question, they are usually characterized as Bingham fluids [46–48].

Sales et al. [11] observed that both lime and lime metakaolin pastes followed the same behavior. Additionally, the authors observed that presence of metakaolin reduced paste initial yield stress from 2 to 1 Pa and viscosity. In case of cementitious pastes that include pozzolanic additions, the behavior that best fits is also that of Herschel-Bulkley [45–47, 49–57].

Zhang et al. [58] studied aerial lime and metakaolin pastes for initial fluidity, macroscopic rheological properties, viscoelasticity and initial dissolution hydration. The authors varied water/binder ratio, the superplasticizer and metakaolin content. The superplasticizer increased fluidity. The pastes fitted Bingham and Herschel-Bulkley model. Also according to authors, variation in viscoelastic characteristics had more influence on water and superplasticizer content than on metakaolin content, so dispersion is more prominent than formation of the network structure of the lime-metakaolin pastes.

#### *5.1.4 Influence of existing commercial additives*

Arizzi and Cultrone [14] analyzed eight lime and metakaolin mortars with varying metakaolin content, proportion of binder: aggregate and different types of additives. The additives did not produce morphological and mineralogical changes in mortars, but they reduced necessary water content and, consequently, reduced porosity and increased resistance. Even so, an increase in metakaolin content stands out among variables analyzed and results in changes in the most significant properties.

A more recent study also shows lime mortars behavior with metakaolin and with nanosilica, another pozzolan with potential use in lime mortars [42]. The authors also incorporated three additives: an adhesion improver, waterproofing agent and viscosity modifier. The objective was to evaluate the influence of combination on several properties, such as fluidity, hardening time, adhesion, cracking, porous structure, resistance to freezing and durability. The general conclusion regarding additives was that nanosilica increased strength and durability while metakaolin improved adhesion and cracking.

## 6. Conclusion

Historical buildings require conservation and restoration services with compatible mortars, that is, similar to original materials. Prior to the emergence of Portland cement, the widespread binder was lime. In addition to lime, historical records also point to the use of pozzolans in old buildings, which justifies their use nowadays, along with the improvement in properties, such as hardening time and durability. Pozzolans also influence the setting and hardening mechanisms, the resulting chemical compounds and rheological aspects.

However, for its current use, studies have been carried out, since much knowledge of practice and construction techniques was lost with the advent of Portland cement and the absence of old records. In addition, the inclusion of pozzolans in lime mortars implies particularities in setting and hardening process, being influenced by the type of pozzolan used and its pozzolanic activity, replacement content in mortar, environmental conditions of application and curing, among others.

Therefore, this chapter highlighted the factors that influence lime mortars and how they respond to inclusion of pozzolans in terms of setting and hardening mechanisms, mechanical performance, formed hydration products and rheological aspects.

## Acknowledgements

The authors are grateful to the CAPES Brazilian Funding (Finance Code 001) and Federal University of Paraiba, Brazil.

## Conflict of interest

The authors declare no conflict of interest.

## Author details

Leane Priscilla Bonfim Sales[1], Aline Figueiredo da Nóbrega[1*], Iranilza Costa da Silva[2], Ana Cecília Vieira da Nóbrega[3], Arnaldo Manoel Pereira Carneiro[2], Fabiola Luana Maia Rocha[1] and Diego de Paiva Bezerra[1]

1 Federal University of Paraíba, João Pessoa, Brazil

2 Federal University of Pernambuco, Recife, Brazil

3 Federal University of Rio Grande do Norte, Natal, Brazil

*Address all correspondence to: alinefnobrega@hotmail.com

## References

[1] Veiga M, do R. Argamassas de cal para conservação e reabilitação de edifícios: conhecimento consolidado e necessidades de investigação. Ambient Construído. 2018;**18**(4):85-96

[2] Papayianni I, Stefanidou M. Strength-porosity relationships in lime-pozzolan mortars. Construction and Building Materials. 2006;**20**(9): 700-705

[3] Magalhães A, Veiga R. Caracterización física y mecánica de los morteros antiguos. Aplicación a la evaluación del estado de conservación. Materiales de Construcción. 2009;**59**(295):61-77

[4] Sousa AKD. Argamassas do grupo escolar Augusto Severo/ RN: Caracterização e incidência de manifestações patológicas. Universidade Federal do Rio Grande do Norte; 2014. Available from: http://www.repositorio.ufrn.br:8080/jspui/bitstream/123456789/14858/1/AdlaKDS_DISSERT.pdf

[5] Gameiro A, Santos Silva A, Veiga R, Velosa A. Hydration products of lime-metakaolin pastes at ambient temperature with ageing. Thermochimica Acta. 2012;**535**:36-41. DOI: 10.1016/j.tca.2012.02.013

[6] Gameiro A, Santos Silva A, Faria P, Grilo J, Branco T, Veiga R, et al. Physical and chemical assessment of lime-metakaolin mortars: Influence of binder: Aggregate ratio. Cement and Concrete Composites. 2014;**45**:264-271. DOI: 10.1016/j.cemconcomp.2013.06.010

[7] Seabra MP, Paiva H, Labrincha JA, Ferreira VM. Admixtures effect on fresh state properties of aerial lime based mortars. Construction and Building Materials. 2009;**23**(2):1147-1153. DOI: 10.1016/j.conbuildmat.2008.06.008

[8] Nežerka V, Slížková Z, Tesárek P, Plachý T, Frankeová D, Petráňová V. Comprehensive study on mechanical properties of lime-based pastes with additions of metakaolin and brick dust. Cement and Concrete Research. 2014;**64**:17-29

[9] Arizzi A, Cultrone G. Comparing the pozzolanic activity of aerial lime mortars made with metakaolin and fluid catalytic cracking catalyst residue: A petrographic and physical-mechanical study. Construction and Building Materials. 2018;**184**:382-390

[10] de AFN A, Struble LJ, Carneiro AMP. Microstructural characteristics of lime-pozzolan pastes made from kaolin production wastes. Materials and Structures/Materiaux et Constructions. 2015;**48**(7):2123-2132

[11] Sales LPB, de Queiroz MGC, da Nóbrega AF, da Nóbrega ACV, de Souza JJN, Carneiro AMP. Study of rheological properties of lime–metakaolin slurries. Applied Clay Science. 2021;**215**

[12] Coelho AZ, Pacheco TF, Said J. A Cal na Construção. Guimarães, Portugal: TecMinho; 2009

[13] Aggelakopoulou E, Bakolas A, Moropoulou A. Properties of lime-metakolin mortars for the restoration of historic masonries. Applied Clay Science. 2011;**53**(1):15-19. DOI: 10.1016/j.clay.2011.04.005

[14] Arizzi A, Cultrone G. Aerial lime-based mortars blended with a pozzolanic additive and different admixtures: A

mineralogical, textural and physical-mechanical study. Construction and Building Materials. 2012;**31**:135-143. DOI: 10.1016/j.conbuildmat.2011.12.069

[15] Grilo J, Santos Silva A, Faria P, Gameiro A, Veiga R, Velosa A. Mechanical and mineralogical properties of natural hydraulic lime-metakaolin mortars in different curing conditions. Construction and Building Materials. 2014;**51**:287-294. DOI: 10.1016/j.conbuildmat.2013.10.045

[16] Silva AS, Gameiro A, Grilo J, Veiga R, Velosa A. Long-term behavior of lime-metakaolin pastes at ambient temperature and humid curing condition. Applied Clay Science. 2014;**88-89**:49-55. DOI: 10.1016/j.clay.2013.12.016

[17] Do Rosário Veiga M, Fragata A, Velosa AL, Magalhães AC, Margalha G. Lime-based mortars: Viability for use as substitution renders in historical buildings. International Journal of Architectural Heritage. 2010;**4**(2):177-195

[18] Elsen J, Van Balen K, Mertens G. Hydraulicity in historic lime mortars: A review. In: Válek J, Hughes J, Groot C, editors. Historic Mortars. RILEM Bookseries. Vol. 7. Dordrecht: Springer; 2012. DOI: 10.1007/978-94-007-4635-0_10

[19] Santiago CC. Argamassas Tradicionais de Cal. Vol. 1. EDUFBA; 2007. p. 202

[20] Veiga R. Air lime mortars: What else do we need to know to apply them in conservation and rehabilitation interventions? A review. Materials and Structures/Materiaux et Constructions. 2017;**157**:132-140

[21] Veiga MDR. As argamassas na conservação. In: 1as Jornadas Eng Civ da Univ Aveiro Avaliação e Reabil das Construções Exist. 2003. pp. 1-22

[22] Magalhães AC. Patologia de rebocos antigos. Cad Edifícios. 2002;**1**:1-15

[23] Sec A, Civil E. Ana Luísa Pinheiro Argamassas de cal Com Pozolanas Para Lomelino Velosa Revestimento de Paredes Antigas. Aveiro, Portugal: Universidade de Aveiro; 2006

[24] Borges C, Santos Silva A, Veiga R. Durability of ancient lime mortars in humid environment. Construction and Building Materials. 2014;**66**:606-620. DOI: 10.1016/j.conbuildmat.2014.05.019

[25] Santos AR, do MR V, Santos Silva A, de Brito J, Álvarez JI. Evolution of the microstructure of lime based mortars and influence on the mechanical behaviour: The role of the aggregates. Construction and Building Materials. 2018;**187**:907-922. DOI: 10.1016/j.conbuildmat.2018.07.223

[26] Veiga MR, Velosa A, Magalhães A. Experimental applications of mortars with pozzolanic additions: Characterization and performance evaluation. Construction and Building Materials. 2009;**23**(1):318-327

[27] Vinagre AIB. Estudo Comparativo Entre Argamassas de Cal Calcítica e de Cal Dolomítica: Características Mecânicas, Físicas, Mineralógicas e Microestruturais. Vol. 1. Universidade De Évora; 2015. p. 203

[28] Instituto Português da Qualidade. NP EN 459-1 - Cal de Construção - Definições, Especificações e Critérios de Conformidade; 2011. pp. 1-50

[29] Zhang D, Zhao J, Wang D, Xu C, Zhai M, Ma X. Comparative study on the properties of three hydraulic lime

mortar systems: Natural hydraulic lime mortar, cement-aerial lime-based mortar and slag-aerial lime-based mortar. Construction and Building Materials. 2018;**186**:42-52. DOI: 10.1016/j.conbuildmat.2018.07.053

[30] Callebaut K, Elsen J, Van Balen K, Viaene W. Nineteenth century hydraulic restoration mortars in the Saint Michael's church (Leuven, Belgium): Natural hydraulic lime or cement? Cement and Concrete Research. 2001;**31**(3):397-403

[31] Nogueira R, Ferreira Pinto AP, Gomes A. Artificial ageing by salt crystallization: Test protocol and salt distribution patterns in lime-based rendering mortars. Journal of Cultural Heritage. 2020;**45**:180-192. DOI: 10.1016/j.culher.2020.01.013

[32] Bakolas A, Aggelakopoulou E, Moropoulou A, Anagnostopoulou S. Evaluation of pozzolanic activity and physicomechanical characteristics in metakaolin-lime pastes. Journal of Thermal Analysis and Calorimetry. 2006;**84**(1):157-163

[33] de AFN A. Estudo do resíduo de caulim em argamassas a base de cal quanto às suas propriedades frescas. Endurecidas E Microestruturais. 2012:229

[34] Alvarez JI, Veiga R, Martínez-Ramírez S, Secco M, Faria P, Maravelaki PN, et al. RILEM TC 277-LHS report: A review on the mechanisms of setting and hardening of lime-based binding systems. Materials and Structures/Materiaux et Constructions. 2021;**54**(2):1-30

[35] Moorehead DR. Cementation by the carbonation of hydrated lime. Cement and Concrete Research. 1986;**16**:700-708

[36] Moropoulou A, Bakolas A, Moundoulas P, Aggelakopoulou E, Anagnostopoulou S. Strength development and lime reaction in mortars for repairing historic masonries. Cement and Concrete Composites. 2005;**27**(2):289-294

[37] Shi C, Day RL. Pozzolanic reaction in the presence of chemical activators: Part II. Reaction products and mechanism. Cement and Concrete Research. 2000;**30**(4):607-613

[38] De Azerêdo AFN, Carneiro AMP, De Azerêdo GA, Sardela M. Hardened properties of lime based mortars produced from kaolin wastes. Key Engineering Materials. 2014;**600**:282-296

[39] Velosa AL, Rocha F, Veiga R. Influence of chemical and mineralogical composition of. Acta Geodynamics et Geomaterialia. 2009;**6**(1):121-126. Available from: https://www.researchgate.net/publication/235960109_Influence_of_chemical_and_mineralogical_composition_of_metakaolin_on_mortar_characteristics

[40] Matias G, Faria P, Torres I. Lime mortars with ceramic wastes: Characterization of components and their influence on the mechanical behaviour. Construction and Building Materials. 2014;**73**:523-534. DOI: 10.1016/j.conbuildmat.2014.09.108

[41] Walker R, Pavía S. Physical properties and reactivity of pozzolans, and their influence on the properties of lime-pozzolan pastes. Materials and Structures/Materiaux et Constructions. 2011;**44**(6):1139-1150

[42] González-Sánchez JF, Fernández JM, Navarro-Blasco AJI. Improving lime-based rendering mortars with admixtures. Construction and Building Materials. 2021;**271**

[43] Moropoulou A, Cakmak A, Labropoulos KC, Van Grieken R, Torfs K. Accelerated microstructural evolution of a calcium-silicate-hydrate (C-S-H) phase in pozzolanic pastes using fine siliceous sources: Comparison with historic pozzolanic mortars. Cement and Concrete Research. 2004;**34**(1):1-6

[44] Fourmentin M, Ovarlez G, Faure P, Peter U, Lesueur D, Daviller D, et al. Rheology of lime paste—A comparison with cement paste. Rheologica Acta. 2015;**54**(7):647-656

[45] Betioli AM, John VM, Jean P, Gleize P, Pileggi RG. Caracterização Reológica de Pasta Cimentícia: Associação de Técnicas Complementares. Porto Alegre, Brazil: Ambiente Construído (JOURNAL); 2009. pp. 37-48

[46] Banfill PFG. Rheological methods for assessing the flow properties of mortar and related materials. Construction and Building Materials. 1994;**8**(1):43-50

[47] Struble L. Rheological changes associated with setting of cement paste. Advanced Cement Based Materials. 1995;**2**(6):224-230

[48] Azeredo G, Morel JC, Lamarque CH. Applicability of rheometers to characterizing earth mortar behavior. Part I: Experimental device and validation. Materials and Structures/Materiaux et Constructions. 2008;**41**(8):1465-1472

[49] Ferrara L, Cremonesi M, Tregger N, Frangi A, Shah SP. On the identification of rheological properties of cement suspensions: Rheometry, computational fluid dynamics modeling and field test measurements. Cement and Concrete Research. 2012;**42**(8):1134-1146. DOI: 10.1016/j.cemconres.2012.05.007

[50] Giordani C, Masuero AB. Blended mortars: Influence of the constituents and proportioning in the fresh state. Construction and Building Materials. 2019;**210**:574-587. DOI: 10.1016/j.conbuildmat.2019.02.077

[51] Papo A, Caufin B. A study of the hydration process of cement pastes by means of oscillatory rheological techniques. Cement and Concrete Research. 1991;**21**(c):1111-1117

[52] Struble LJ, Schultz MA. Using creep and recovery to study flow behavior of fresh cement paste. Cement and Concrete Research. 1993;**23**(6):1369-1379

[53] Banfill P, Frias M. Rheology and conduction calorimetry of cement modified with calcined paper sludge. Cement and Concrete Research. 2007;**37**(2):184-190

[54] Banfill PFG, Rodríguez O, Sánchez de Rojas MI, Frías M. Effect of activation conditions of a kaolinite based waste on rheology of blended cement pastes. Cement and Concrete Research. 2009;**39**(10):843-848

[55] Janotka I, Puertas F, Palacios M, Kuliffayová M, Varga C. Metakaolin sand-blended-cement pastes: Rheology, hydration process and mechanical properties. Construction and Building Materials. 2010;**24**(5):791-802. DOI: 10.1016/j.conbuildmat.2009.10.028

[56] Mansour SM, Abadlia MT, Bekkour K, Messaoudene I. Improvement of rheological behaviour of cement pastes by incorporating metakaolin. European Journal of Scientific Research. 2010;**42**(3):442-452

[57] Safi B, Benmounah A, Saidi M. Rheology and zeta potential of cement pastes containing calcined silt and ground granulated blast-furnace

slag. Materiales de Construcción. 2011;**61**(303):353-370

[58] Zhang D, Xu D, Liu H, Wang Y, Wang H, Wang J, et al. Effect of water to binder ratio, polycarboxylate superplasticizer and metakaolin dosages on rheological and viscoelastic properties of fresh metakaolin-air lime pastes. Journal of Building Engineering. 2022;**62**(September):105351. DOI: 10.1016/j.jobe.2022.105351

Chapter 9

# Warm/Cool Feeling Characteristics of Ceramic-Incorporated Fabrics: Process, Characterization, and Thermal Properties

*Seung Jin Kim and Hyun Ah Kim*

## Abstract

Recently, high-performance functional textile goods have been commercialized using various ceramic nanopowders, such as ZrC, $Al_2O_3$, $SiO_2$, ZnO, ATO, and $TiO_2$, embedded in the yarns and fabrics. This study examines the warm-cool feeling characteristics of ceramic-incorporated fabrics with their process, characterization, and thermal characteristics. This topic is divided as follows: review (introduction), preparation (experimental), characterization with thermal property (results and discussion), and summary. As a review, heat release and storage properties of the various ceramic-embedded fabrics are introduced and multifunctional properties of different ceramic-embedded fabrics such as UV-cut and anti-static with thermal wear comfort are reviewed with types of ceramic particles embedded in the yarns through the literature published up to now. In the text, warm-cool feelings of ceramic-embedded fabrics prepared in this study are compared in terms of heat flow rate (Qmax) and thermal insulation value (TIV) and summarized with dye-affinity and color-fastness of the ceramic-embedded fabrics. Finally, the future prospect for functional to textiles treated with ceramic materials is proposed in the fields of water repellence, anti-bacteria, flame retardation, anti-static, and UV protection.

**Keywords:** coolness, warm feeling, ceramic, heat flow rate (Qmax), thermal insulation value (TIV), FIR, dye-affinity, color fastness

## 1. Introduction

Heat storage and retention warm-up textile materials have been used in sportswear and lingerie with eco-friendly and environmentally functional textile materials. Many commercialized warm-up textile materials were developed by Japanese companies, such as Unitica [1], Kuraray [2], Mitsubishi-rayon [3], Toray [4], and KB seiren [5]. Many studies and technical reports for this area have been provided despite the confidentiality related to technical information by the manufacturing company [6–9]. Improving heat storage and retention of the warm-up textile materials in the company mainly involves three types of technology: micro-multilayer fiber with high hollowness [10, 11], heat release/storage by ceramic-embedded technology, and heat development using a phase change material (PCM) [12–14]. Of these technologies, ceramic-embedded technology has attracted considerable attention because of its appearance in the market by many Japanese textile companies.

On the other hand, many studies [15–19] related to their interactions between the heat released from the ceramic particles in the yarns and infrared radiation have been carried out using a range of ceramic particles, such as ZrC, $Al_2O_3$, and $SiO_2$. Negishi and Kikuchi [15] reported the heat-absorbing property of ZrC particles and that they reflected far-infrared (FIR) radiation. Furuta et al. [16] examined the heat storage and moisture permeability of the ZrC-embedded PET fabric. Bahng and Lee [17] reported the heat-generating and rapid moisture absorption/drying properties of the $Al_2O_3$-embedded fabrics. Anderson et al. [18, 19] reported increased solar absorption with an increased $TiO_2$ content embedded in the PET filament and presented the thermal outwear made from $TiO_2$-embedded fabric designed for cold-weather applications. Kim and Kim [20, 21] examined the thermal properties and wear comfort of the ZrC-embedded PET fabrics.

On the other hand, some studies [22–25] related to thermal and wear comfort were performed using regular and special fibers with a change in the yarn and fabric structural parameters. Matsudaira [22] examined the moisture and heat transmission of the eight types of fabric specimens made from different grooved hollow fibers. They reported that the thermal conductivity of the fabric specimens differed according to the cross-sectional shape of the grooved hollow fibers, and the maximum heat flow (Qmax) of the fabric specimens was primarily affected by the contacted surface area between the fibers in the yarn. Furthermore, the thermal insulation value (TIV) of the fabric specimens was strongly dependent on the cross-sectional shape of the grooved hollow fibers. Onofrei et al. [23] examined the influence of knitted fabrics structure on the thermal and moisture properties using Coolmax® and Outlast®. They reported the importance of a knitted fabric structure suitable for summer and winter sportswear by analyzing their heat and moisture characteristics. Supuren et al. [24] examined warm-cool characteristics through the moisture management and thermal absorptivity of double-face knitted fabrics used in sports/activewear using a cotton/polypropylene (PP) blend yarns. Majumdar et al. [25] examined the thermal properties in terms of thermal conductivity, wet thermal resistance, and air permeability of knitted fabrics made from cotton and regenerated bamboo cellulosic fibers. They reported the thermal wear comfort according to the knitted fabric structure and pattern. In addition to these studies, several studies have measured the warm-cool feeling characteristics [26–28] and clothing temperature change using a phase change material [29–31]. However, few studies have examined the thermal wear comfort with the dyeability and color fastness of Nylon/PP sea and island warm-up knitted fabrics incorporated with fine ceramic powders.

On the other hand, high functional fibers have become applicable to intelligent textile materials, where self-regulating control of cold and hot weather is possible by the release and absorption of heat [6, 12]. Among them, cool textile materials used in hot weather are made from phase change materials (PCM) [6, 13, 14] and moisture-responded transformable (MRT) fiber [14, 32], which are called intelligent textiles or smart textiles. Recently, Mather [33] proposed that PCM, shape memory polymers (SMPs), and breathable fabrics, including thermochromism and photochromism, could apply to intelligent textiles. The MRT fiber was developed by Teijin in Japan, which has been commercialized to knitted fabrics for sportswear [32]. MRT fibers are expanded after moisture absorption like the hygral expansion (HE) of wool fibers.

Kim and Kim [32] examined the physical properties of Huvis elastic fiber (HEF) developed by Huvis Co. Ltd. in Korea, which is similar to the MRT fiber of Teijin in Japan. Many studies [24, 34–43] focused on improving the coolness feeling characteristics of sweating have been carried out using fabric manufacturing technology by combining hydrophobic and hydrophilic yarns. Of these studies, D. Mikucioniene et al. [34] examined heat transfer characteristics of the water absorption and fast-drying knitted fabrics made from cotton/synthetic composite yarns. M. Piraee et al. [35] reported the thermal characteristics of 13 acrylic knitted fabrics with different stitch length. M. Yanilmaz [36] investigated wear comfort with absorption and drying properties of acrylic knitted fabrics with different structures. Chen et al. [38–40] reported the extent of the coolness of nylon and PET knitted fabrics using initial absorption rate and wicking rate measured from biomimetics of branching structure. Fan et al. [41] and Sarkar et al. [42, 43] conducted intensive studies to improve the moisture absorption of multilayered fabric on plant-based fabric structures.

On the other hand, commercially applied coolness textile technologies include two types of technology: one is perspiration absorption and fast-dry fiber material with a noncircular cross-sectional shape, Coolmax®, as a well-known brand of Dupont (Invista). In particular, perspiration absorption and fast-drying properties (coolness characteristics) of fabrics using Coolmax® and Outlast® have been examined by some textile scientists [23, 44–47]. Onofrei et al. [23] examined the influence of knitted fabric structure on the thermal and moisture properties of knitted fabrics using Coolmax®. Kim and Kim [44–47] examined the wicking and drying properties as coolness characteristics and heat retention rate as warmness characteristics of woven fabrics made from PET and PP composite yarns with Coolmax®/bamboo/Tencel fibers. They reported that the Coolmax®/Tencel sheath/core fabrics exhibited superior wicking and drying properties applicable to coolness fabrics for summer outdoor clothing.

Another technology uses ceramic-incorporated hydrophilic PET fiber material with superior heat transfer properties using ethylene vinyl alcohol (EVOH) as a hydrophilic heat transfer material, which was commercialized in Japanese companies, such as Kuraray and Komatsu-seiren. Sophista®, made in Kuraray is manufactured using composite spinning technology with three types of yarn cross-sections: sheath/core, multilayer, and multi-hollow with hydrophilic heat transfer material, EVOH, which enables a cool feel to the wearer while it contacts the human skin. Aqusia® made in Komatsu-seiren was commercialized using bicomponent spinning with nylon with ceramic powders and hydrophilic PET. It is composed of a sheath filled with ceramic-incorporated nylon and a core with hydrophilic and high heat transfer PET. On the other hand, despite the many fiber manufacturing companies providing considerable technical information as a commercial base, various physical properties, including the coolness feel of the hydrophilic and heat transfer PET, are less known for confidentiality reasons. Recently, Huvis Co. Ltd. in Korea developed a ceramic-embedded hydrophilic PET coolness filament with noncircular and different shaped cross-sections, but there are no reports of the detailed coolness feeling and dyeability data according to the dyeing processing factors in terms of the dyeing temperature and time.

Therefore, this study examined warm/cool feeling characteristics of ceramic incorporated fabrics, and which is composed of two part, first part examined the thermal wear comfort in terms of maximum heat flow (Qmax) and thermal insulation value (TIV) of the warm-up knitted fabric specimens made from Nylon/PP sea and island composite filaments incorporated with fine ceramic particles. Second part examined the physical properties of the ceramic-embedded hydrophilic PET coolness filament developed by Huvis Ltd. Co. in Korea. The coolness feeling of the knitted fabric made from the ceramic-embedded coolness filament was compared with that of control knitted fabrics made from hydrophobic PET coolness and regular PET filaments. In particular, the effect of ceramic powders incorporated in the coolness filament to the cool feeling was examined. In addition, the dye-affinity, color difference, and color fastness to the washing of hydrophilic PET coolness fabrics with noncircular and different shaped yarn cross-sections were compared with dyeing process factors, such as different dyeing temperatures and times, and discussed with those of the control fabrics, such as hydrophobic PET coolness and regular PET fabrics.

## 2. Literature review on application of ceramics to textiles goods

Improving the warm feel of textiles mainly involves three types of technology: heat of wetting by water absorption, heat release using phase change material (PCM), and heat storage/release by ceramic-incorporated technology. On the other hand, cool feel technologies of textiles are classified into four types: perspiration absorption and fast-drying PET with a noncircular cross-sectional shape (Coolmax®) and nylon high hollow fiber (Wincall®) using conjugated spinning, hydrophilic PET fiber with superior heat transfer property using ethylene vinyl alcohol (EVOH) by bicomponent (sheath/core) spinning, PCM material, and moisture-responded transformable (MRT) fiber technologies. Of these technologies, hydrophilic PET fiber with EVOH using a bicomponent spinning is composed of a sheath filled with ceramic powder and a core with hydrophilic and EVOH PET polymer. As mentioned above, ceramic powders are applied to both warm and cool-feeling textiles.

Therefore, many studies related to warm/cool feeling textiles have been explored using ceramic powders with many researches to improve UV protection, anti-static, and anti-bacteria properties. Recently, high-performance functional textile goods have been commercialized using various ceramic nanopowders, such as ZrC, $Al_2O_3$, $SiO_2$, ZnO, ATO, and $TiO_2$, embedded in the yarns and fabrics. Among several methods treated with ceramic powders to the textiles, coating is a common technique used to apply nanopowders onto textiles, however, the coating has some issue in washing durability according to the repeated washing and laundering of the treated textiles. In recent, a new method by novel scheme, not a conventional one such as coating method was developed. Accordingly, recent studies related to heat release and storage by thermal radiation, including UV protection and anti-static properties, using various ceramic particles are critically reviewed in this literature survey. In early years, scientific studies [15–19] related to their interactions between heat released from various ceramic particles embedded in the yarns and infrared radiation were conducted using various ceramic particles, such as ZrC [15, 16], $Al_2O_3$, $SiO_2$ [17], and $TiO_2$ [18, 19]. More recently, Kim and Kim [20, 21, 48–50] conducted an intensive study on the heat release/storage and thermal wear comfort of ZrC/$Al_2O_3$-embedded PET fabrics. They [20, 21] examined FIR emission characteristics of ZrC-embedded knitted fabrics and reported the thermal wear comfort of the ZrC-embedded fabrics ***via*** thermal manikin experiment. They concluded that ZrC absorbs the heat emitted from the human body or reflects the FIR radiation, which prevents the heat in the clothing and human body from flowing out.

In addition, Kim and Kim [21] examined heat release/storage properties and thermal wear comfort of $Al_2O_3$/graphite and ZrC/graphite-embedded fabrics. They reported that the heat release characteristics of the ZrC/graphite-embedded fabrics were superior to those of the $Al_2O_3$/graphite-embedded fabrics. In addition, various wear comfort properties of the fabrics using thermal manikin and wearer trials with human subjects were examined and reported that the wear comfort properties of the ZrC/graphite-embedded fabric were superior to those of the $Al_2O_3$/graphite-embedded fabric. In further studies, Kim [48] examined the thermal radiation and wear comfort properties of the $Al_2O_3$/graphite-embedded fabrics according to the embedded yarn structure distributed by ceramic particles in the yarns. Moisture absorption and drying properties using a moisture management tester (MMT) and thermal insulation property by a KES-F7 system were measured and compared with two ceramic-embedded yarn structures (sheath/core and dispersed yarn). In particular, Kim [49, 50] examined the dry and wet thermal wear comfort properties of the $Al_2O_3$/graphite-embedded fabrics for cold weather protective clothing according to ceramic particles distributed yarn structures using a sweating thermal manikin apparatus [49]. Thermal insulation (Clo value) using a thermal manikin of the $Al_2O_3$/graphite-embedded fabrics was compared with thermal radiation (FIR emissivity and maximum surface temperature) in terms of yarn structure such as sheath/core, dispersed, and regular PET yarns [50].

On the other hand, many studies [51–62] using various ceramic particles to improve UV protection and anti-static characteristics of the fabrics have been conducted with fabrics coated with different ceramic powders. Of these studies, some studies [51, 52] examined the increase of the effectiveness of UV protection by coating treatment of $TiO_2$ and others [53–55] reported more efficient UV protection of ZnO-coated fabric than conventional one. $TiO_2$ and ZnO among various ceramic particles were used as UV blockers, which were more efficient at absorbing and scattering UV radiation than the other ceramic particles. In contrast, ATO (antimony tin oxide) and ZnO particles provide an anti-static property with electrically conductive characteristics [56–62], which helps to effectively dissipate the static charge accumulated on the fabrics coated with ZnO [59, 62] and ATO nanopowders [56–58, 60, 61]. In particular, ATO particles have a thermal insulation property with a heat shielding effect [58]. However, most of the previous studies given by the coating treatment have limitation to the durability of the function during wearing clothing and laundering. Hence, functional yarns and fabrics produced from new scheme, not coating have been required.

Most recently, Kim [63, 64] reported UV protection and anti-static characteristics of the ATO/$TiO_2$/$Al_2O_3$-embedded PET fabrics produced from new scheme, that is, bicomponent spinning method, and showed the applicability of these yarns to workwear protective clothing because of their superior UV protection and anti-static characteristics. By the way, the hazard posed by static electricity is heightened considerably in cold and dry environments in cold weather regions, particularly, oil and gas industries are located, where the workwear protective clothing requires superior thermal wear comfort with the anti-static property. In addition, apart from dramatically reducing exposure to the sun to the workman exposed to sunlight, the most frequently recommended form of UV protection is the use of UV protective clothing. Accordingly, workwear protective clothing wearing in winter and cold weather regions requires multifunctional high performance such as anti-static and UV cut with superior thermal wear comfort.

In recent, Kim [65, 66] examined the UV protection and anti-static properties with thermal radiation of the $Al_2O_3$/ZnO/ZrC/ATO-embedded fabrics to investigate their properties according to the mixing of ceramic particles embedded in the yarns. They reported that ZnO/ZrC and ZnO/ATO-embedded fabric exhibited superior UV protection due to excellent UV protection characteristics of the ZnO ceramic particles. ATO-embedded fabric showed superior anti-static properties with lower rub-static voltage and surface electrical resistivity (SER). Based on multifunctional high-performance protective fabric, it was found that ZnO/ATO-embedded fabric is of practical use for engineering workwear protective clothing with superior UV protection factor (UPF), excellent anti-static property, and relatively good thermal radiation. In addition, ZnO/ZrC-embedded fabric was expected to be applicable to workwear protective clothing wearing in cold weather regions because of excellent thermal radiation, UPF, and relatively good anti-static property.

## 3. Experimental

### 3.1 Specimen preparation

#### *3.1.1 Yarn specimens*

The Nylon/PP composite filament was spun as sea and island-type partially oriented yarns (POY) on the conjugated spinning machine installed in Huvis Co. Ltd. (Suwon, Korea). Ceramic nanopowders (ZrC, 0.6 wt.%) were embedded in the core part with PP polymer on the bicomponent spinning machine. Ceramic-incorporated Nylon/PP warm-up POY was texturized on a texturing machine (AIKI, Japan), which was used as a warm-up yarn specimen (W-1) to make a warm-up knitted fabric specimen. In addition, two types of ceramic-incorporated PET (W-2 and 3) and one regular PET (W-4) yarn were used as the control yarns to make the knitted fabric specimens, which were compared with the ceramic-incorporated nylon/PP warm-up knitted fabric.

On the other hand, the yarn specimen used as a coolness yarn was developed and spun on a pilot spinning machine in Huvis Co. Ltd. (Suwon, Korea). The developed yarn specimen (C-1) was a hydrophilic ceramic-embedded ($TiO_2$, 0.3 wt.%) PET coolness filament with noncircular and different-shaped cross-sections. One control yarn specimen (C-2) was a hydrophobic PET coolness filament, and the other control yarn specimen (C-3) was a regular PET filament. The hydrophilic PET polymer was first polymerized to impart hydrophilicity by adding intermittently polyol molecules between the molecular chains of terephthalic acid (TPA) and ethylene glycol (EG). A noncircular and different-shaped cross-section was imparted in the spinning process. The ceramic nanopowder was incorporated into this process to give a coolness feel. **Table 1** lists the specification of the yarn specimens.

#### *3.1.2 Knitted fabric specimens*

Warm-up knitted fabric specimens were fabricated using four types of yarn specimens on a circular knitting machine (Double knitting m/c, Geumyoung, Daegu, Korea). The specifications of the knitting machine for warm-up knitted fabrics were as follows: 30-inch diameter, 18 gauge, 2640 stitch, and 18 rpm speed, and 28 gauge for coolness fabric specimens, was used with 15 rpm speed. The pattern of the warm-up knitted fabric specimens was a double jersey with the loop side used by four types of yarn specimens (**Table 1**) and the float (back) side with covering yarn composed of 150d/96f PET DTY and spandex 20d.

The pattern of the coolness knitted fabric specimens was also double knit. **Table 2** lists the patterns of the warm-up and coolness knitted fabric specimens.

| Specimens | Specification | Characteristics | Remark |
|---|---|---|---|
| W-1 | Nylon/PP $75^d/36^f$ | Sea/island composite yarn | Developed |
| W-2 | PET $75^d/72^f$ | Nano-ceramic-incorporated PET | control |
| W-3 | PET $75^d/72^f$ | Ceramic-incorporated PET | control |
| W-4 | PET $75^d/72^f$ | Regular PET | control |
| C-1 | PET $75^d/36^f$ | Hydrophilic PET coolness yarn | Developed |
| C-2 | PET $75^d/72^f$ | Hydrophobic PET coolness yarn | control |
| C-3 | PET $75^d/72^f$ | Regular PET yarn | control |

**Table 1.**
*Characteristics of the yarn specimens.*

| Warm-up knitted fabric specimen | Characteristics (Design) | | | | | | | | | | |
|---|---|---|---|---|---|---|---|---|---|---|---|
| **Pattern (double knit & fancy structure)** | **Cam sequence** | | | | | | | | | | |
| | **Yarn sequence** | | | **a** | **b** | **c** | **d** | **e** | **f** | | |
| | Dial | | B | ∪ | ∨ | — | — | ∨ | — | | |
| | | | A | — | ∨ | — | ∪ | ∨ | — | | |
| | Cylinder | | A | — | — | ∧ | ∩ | — | ∧ | | |
| | | | B | | — | ∧ | — | — | ∧ | | |
| | **Stitch sequence** | | | | | | | | | | |
| | Dial | | A | | B | | A | | B | | |
| | Cylinder | | A | | B | | A | | B | | |
| **Coolness knitted fabric specimens** | **Characteristics (Design)** | | | | | | | | | | |
| **Pattern (double knit & interlock single pique structure)** | **Cam sequence** | | | | | | | | | | |
| | Yarn sequence | | | **a** | **b** | **c** | **d** | **e** | **f** | **g** | **h** |
| | | Dial | B | ∪ | ∨ | ∪ | ∨ | — | ∨ | — | ∨ |
| | | | A | — | ∨ | — | ∨ | ∪ | ∨ | ∪ | ∨ |
| | | Cylinder | A | ∧ | — | ∧ | — | ∧ | — | ∧ | — |
| | | | B | ∧ | — | ∧ | — | ∧ | — | ∧ | — |
| | Stitch sequence | Dial | A | B | A | B | A | B | A | B | |
| | | Cylinder | A | B | A | B | A | B | A | B | |

*Yarn: a, b, d, e: surface yarn; sequence c, f: PET 150d/96f SD: DTY + Spandex 20d. note: ∪, ∩: tuck; ∧, ∨: knit; —: miss.*
*note: ∪: tuck; ∨, ∧: knit; —: miss.*

**Table 2.**
*Pattern and design of the knitted fabrics.*

### *3.1.3 Dyeing and finishing treatments of the knitted specimens*

The Nylon/PP knitted fabric specimen (W-1) was scoured with $Na_2CO_3$, 2 g/ℓ and Sunmorl S-30, 1 g/ℓ at 80°C for 20 minutes in a CPB scouring machine. After scouring, the knitted fabric specimens were dyed with 3% o.w.f. of C.I. acid blue 288 with newborn, 1 g/ℓ, acetic acid PH-4 in a rapid machine. The PET knitted fabric specimens (W-2, −3, and − 4) were dyed with C.I. disperse blue 78 with RM 340, 0.5 g/ℓ, and washed with NaOH, 2 g/ℓ and $Na_2S_2O_4$, 1 g/ℓ at 80°C for 20 minutes in a CPB scouring machine. The dyeing conditions of the warm-up knitted fabric specimens were shown in **Table 3**. The process conditions for coolness knitted fabrics were provided in **Table 3**. The optimal dyeing conditions were determined by changing the dyeing time and temperature for the nylon/PP warm-up and coolness PET knitted fabric specimens. **Table 3** lists the dyeing conditions of the knitted fabric specimens.

### 3.2 Measurement of yarn physical properties

The tenacity and breaking strain of the yarn specimens were measured according to KSK 0416 with a gauge length of 100mm and a crosshead speed of 100mm/min using Testomeric Micro 350 (USA); 20 readings were taken for each specimen. The dry and wet thermal shrinkages were measured using the KSK 0215 measuring method. Cross-sections of the yarn specimens were assessed to determine the ceramic particles incorporated in the yarns using scanning electron microscopy (SEM, S-4300, Hitachi Co., Japan).

### 3.3 Measurement of the thermal properties of the knitted fabric specimens

The thermal wear comfort of the warm-up knitted fabric specimens was measured using a KES-F7 (Thermolabo II, Kato Tech. Co. LTD., Japan) measuring apparatus [67]. The maximum heat flow at the transient state (Qmax, $J/cm^2{\cdot}s$) of the knitted fabric specimens were measured using a KES-F7 system at 20 ± 1°C and 70 ± 5% RH. A 10 cm x 10 cm specimen was prepared and assessed using Eq. (1). An average of five readings for each specimen was reported.

$$Q_{max} = [\int q(t)dt] = \frac{\mathbf{M \cdot C}}{\mathbf{A}} \frac{\mathbf{dT(t)}}{\mathbf{dt}} \mathbf{dt} \quad (1)$$

| Specimens | Dyeing condition | | Temp.increment | M/C |
|---|---|---|---|---|
| | Temp (°C) | Time (min) | | |
| W-1 | 80,90,100 | 30,40,60 | 3 °C/min | IR dyeing m/c |
| W-2 | 110,120,130 | 30,40,60 | | |
| W-3 | 110,120,130 | 30,40,60 | | |
| W-4 | 110,120,130 | 30,40,60 | | |
| C-1 | 110,120,130 | 30,40,60 | • Scouring<br>Soda ash($Na_2CO_3$)2 g/l,<br>Scouring agent(Sunmorl SS-30) 1 g/l, 80°C, 20 min<br>• Dyeing<br>Disperse dye (FORONBLUES–BGL 200–Claiant),(C.I. Disperse Blue 78)<br>dispersing agent(RM340)0.5 g/l<br>• Weight reduction<br>NaOH 2 g/l, $Na_2S_2O_4$ 1 g/l, 80°C, 20 min | |
| C-2 | 110,120,130 | 30,40,60 | | |
| C-3 | 110,120,130 | 30,40,60 | | |

**Table 3.**
*Dyeing conditions of the knitted fabric specimens.*

where q(t) is the heat absorption rate per unit area of the specimen (cal/sec·cm$^2$); A is the area of the plate; M is the mass of the heat plate; C is the specific heat of heat plate (cal/g·°C); and dT(t)/dt is the heat decreasing rate of the heat plate.

Thermal insulation value (TIV) of the knitted fabric specimens was measured as a measure of the heat insulation rate using Eq. (2) [67].

$$\mathrm{TIV} = \left(\mathbf{1} - \frac{\mathbf{b}}{\mathbf{a}}\right) \times \mathbf{100} \tag{2}$$

where a is the heat loss (W) of the plate without a specimen, and b is the heat loss (W) of the plate covered with a specimen.

### 3.4 Measurement of the dye affinity and color fastness of the knitted fabric specimens

The dye affinity characteristics of the knitted fabric specimens were examined at different dyeing temperatures and times. Reflectance (R) of the knitted fabric specimens according to the dyeing temperature and time was measured using a color measuring device (Gretag Macbath, Color-Eye 3100, USA). The dye affinity (K/S) was calculated from R using Eq. (3) [68].

$$\mathrm{K/S} = (1 - \mathrm{R})^2/2\mathrm{R} \tag{3}$$

where R, K, and S are the reflectance, absorption factor, and scattering factor, respectively.

In addition, the color difference (ΔE) was measured, and the color fastness to washing of the knitted fabric specimens was assessed according to the KS K ISO 105-C06.

## 4. Results and discussion

### 4.1 Physical properties of warm-up yarn specimens

The weavability of the yarns is very important for estimating fabric productivity, which can be estimated and compared by the tenacity and breaking strain of yarn. **Table 4** lists the tenacity and breaking strain of the warm-up yarn specimens.

| Specimen | Tenacity (g/d) | | Breaking strain (%) | | Thermal shrinkage (%) | | | |
|---|---|---|---|---|---|---|---|---|
| | Mean | Dev. | Mean | Dev. | Mean(D) | Dev. (D) | Mean(W) | Dev. (W) |
| W-1 | 4.086 | 0.223 | 4.515 | 2.534 | 18.6 | 3.8 | 18.8 | 2.6 |
| W-2 | 4.661 | 0.328 | 1.672 | 1.853 | 18.2 | 1.7 | 16.7 | 6.9 |
| W-3 | 4.056 | 0.303 | 0.600 | 0.811 | 14.7 | 4.7 | 13.6 | 4.3 |
| W-4 | 4.721 | 0.481 | 1.563 | 2.015 | 14.5 | 2.3 | 14.3 | 7.7 |

*Note: dev. = max. – min. D: dry, W: wet.*

**Table 4.**
*Physical properties of warm-up yarn specimens.*

As shown in **Table 4**, the tenacity of the warm-up yarn specimens ranged from 4 to 5 gf/d, which is estimated as the free value of weavability in the weaving process. In addition, the tenacity of nylon/PP filament (W-1) was slightly lower (4 gf/d) than that of the other yarn specimens (W-2, −3, and − 4). Furthermore, W-1, −2, and − 3 exhibited lower tenacity than the regular PET yarn (W-4). This suggests that sea/island yarn and nano-ceramic-incorporated yarn deteriorated yarn tenacity because of the boundary characteristics of the two components between the base polymer in sea/island yarn and nanoparticles incorporated in the yarns. **Figure 1** presents SEM images of the cross-section of yarn specimens, W-1, −2, and − 3. The sea/island yarn cross-section in **Figure 1a** and nanoparticles of ceramics in **Figure 1a–c** are shown as white spots.

As shown in **Table 4**, the breaking strain of the Nylon/PP yarn (W-1) was much higher (4.5%) than that of the other three yarn specimens because of the difference between the nylon and PET yarn characteristics. As shown in **Figures 2**, 36 filaments in yarn specimen 1 were shown, and 72 filaments in yarn specimens 2, 3, and 4 were appeared, respectively. In particular, the breaking strain of yarn specimen 3 showed the lowest value, which was attributed to the larger ceramic particles incorporated in yarn specimen 3 than the other yarn specimens. **Figure 1c** shows that the ceramic particles in yarn specimen 3 are larger than that of yarn specimens 1 and 2 (**Figure 1a, b**), and these particles enable the yarn subjected to stress to be inextensible, resulting in a lower breaking strain.

Estimating the thermal shrinkage of the newly developed yarns is essential for designing relevant dry and wet thermal shrinkages of fabric in the dyeing and finishing processes. Therefore, in this study, the thermal shrinkages of newly developed nylon/PP warm-up composite yarn were compared with the other three yarn specimens. **Table 4** lists the dry and wet thermal shrinkages of the yarn specimens. The dry and wet thermal shrinkages of the nylon/PP warm-up composite yarn (W-1) ranged from 18 to

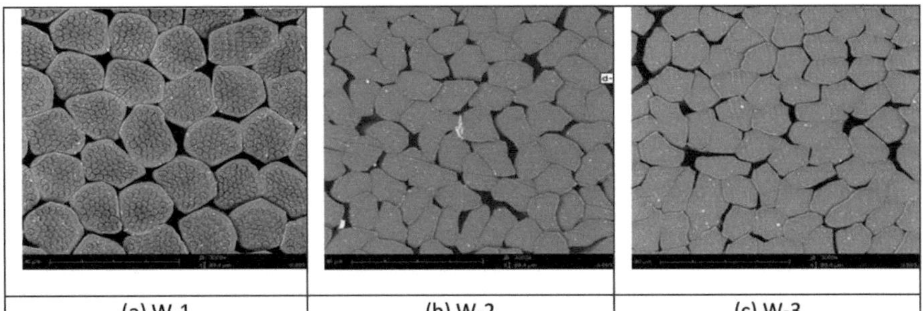


**Figure 1.**
*SEM images of a cross-section of the ceramic particle-incorporated yarns. (a) W-1, (b) W-2, (c) W-3.*

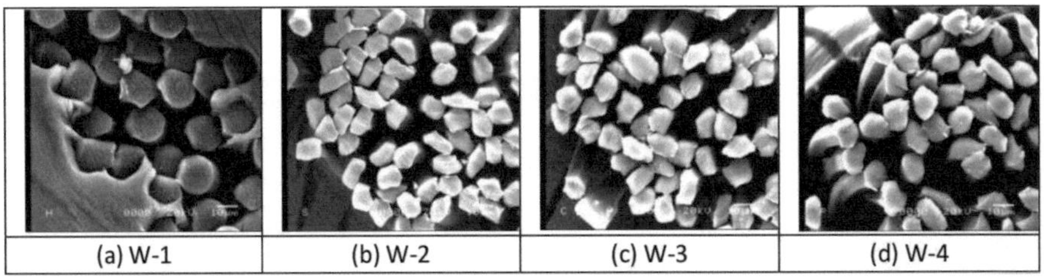


**Figure 2.**
*SEM images of the cross-sections of the yarn specimens (x1000). (a) W-1, (b) W-2, (c) W-3, (d) W-4.*

19%, which were higher than those of specimens 3 and 4, respectively, which was attributed to the higher dry and wet thermal shrinkages of the nylon and PP than those of PET. In addition, the dry and wet thermal shrinkages of the nylon/PP warm-up composite yarn (W-1) and nano-ceramic particle-incorporated PET yarn (W-2) exhibited a similar value. This suggested that nano-ceramic particles incorporated in the yarns imparted thermal stability to the dry and wet thermal deformation.

## 4.2 Thermal wear comfort of the warm-up knitted fabric specimens

### *4.2.1 Contact-warm feeling*

The Qmax (maximum heat flux) was defined as a measure of the warm/cool feeling by contacting fabric measured using the KES-F7 apparatus. **Figure 3** shows the mean of Qmax of the warm-up knitted fabric specimens according to dyeing temperature and time.

The Qmax at the transient state of the nylon/PP knitted fabric (W-1) was lower than that of the ceramic-incorporated and regular PET knitted fabrics (W-2, −3, and − 4). This was attributed to the greater heat emitted from the ceramic particles incorporated in the core of the nylon/PP composite yarns, which prevented heat flow from the human body to outside through the fabric, resulting in a lower Qmax of the nylon/PP knitted fabric specimen even though the thermal conductivity of the nylon is higher than that of the PET. This result was partly due to the lower thermal conductivity of ceramic-incorporated PP than $TiO_2$-incorporated PET [69], which resulted in a lower Qmax of the nylon/PP knitted fabric (W-1) than the ceramic-incorporated PET fabrics (W-2, −3, and − 4).

### *4.2.2 Thermal insulation value (TIV) of the warm-up knitted fabric specimens*

Understanding how the ceramic-incorporated characteristics of nylon/PP sea and island composite yarns affect the TIV of the nylon/PP warm-up knitted fabric is

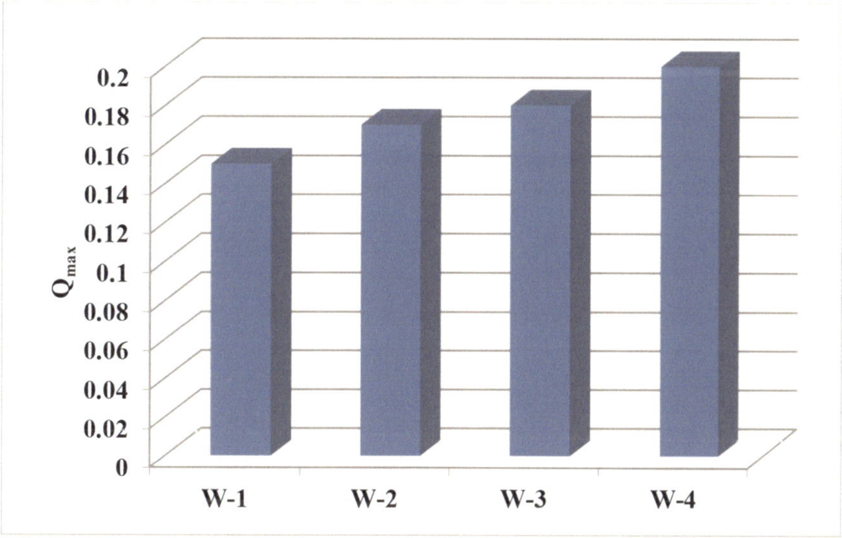


**Figure 3.**
*Qmax of the warm-up knitted fabric specimens.*

essential to evaluating the thermal wear comfort for winter clothing. **Figure 4** presents a diagram of the TIV of the warm-up knitted fabric specimens. The TIV of the nylon/PP warm-up knitted fabric (W-1) was higher than that of the other knitted fabrics (W-2, −3, and − 4), which was attributed to the greater heat emanating from the ceramic particles incorporated in the islands of the nylon/PP sea and island composite yarn than PET yarns and lower thermal conductivity of the PP in the core of the composite yarns than PET yarns, which prohibited heat flow from human skin to the outside through the knitted fabric, resulting in a higher TIV of the nylon/PP warm-up knitted fabrics than the ceramic-incorporated and regular PET fabrics.

The TIV of the nylon/PP warm-up knitted fabric (W-1) was approximately 9.3% higher than that of the ceramic-incorporated PET knitted fabrics (W-2 and -3), highlighting the importance of the ceramic particles embedded in the sea and island yarn structure with a low thermal conductivity of PP incorporated in the island to the TIV of the fabrics.

### 4.3 Dyeability and color fastness of warm-up knitted fabric

#### *4.3.1 Dye affinity characteristics*

The dyeing characteristics of the nylon/PP ceramic-incorporated fabric were compared with those of the ceramic-incorporated PET and regular PET fabrics according to the dyeing temperature and time. **Figure 5** presents the dye affinities (K/S) of the warm-up knitted fabric specimens. The K/S of the nylon/PP knitted fabric (W-1) was lower than that of the ceramic-incorporated PET and regular PET fabrics (W-2, −3, and − 4). This was attributed to the low dye affinity of the PP fibers in the island of the nylon/PP sea and island composite yarns compared to PET and nylon. In addition, the changes in the dye affinity according to the dyeing temperature and time appeared differently between the nylon/PP and three PET fabrics.

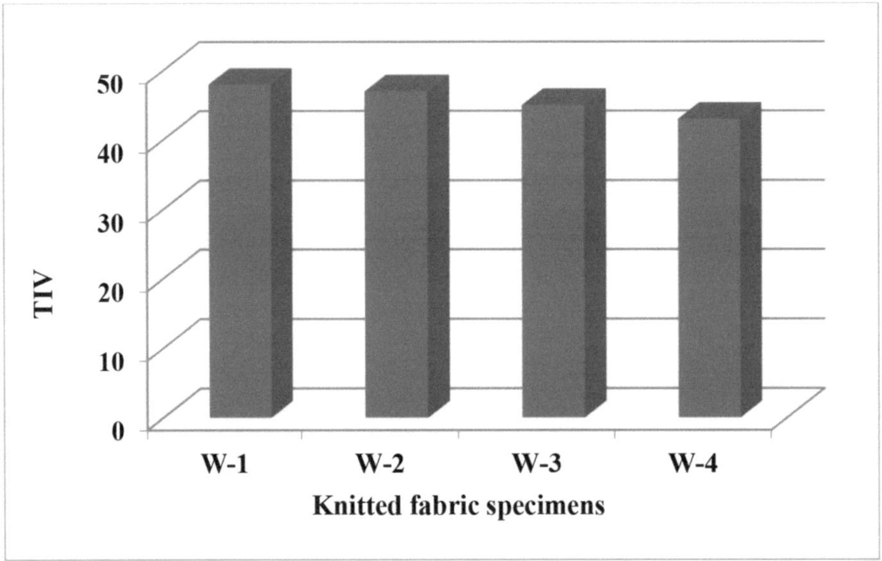


**Figure 4.**
*TIV of the warm-up knitted fabric specimens.*

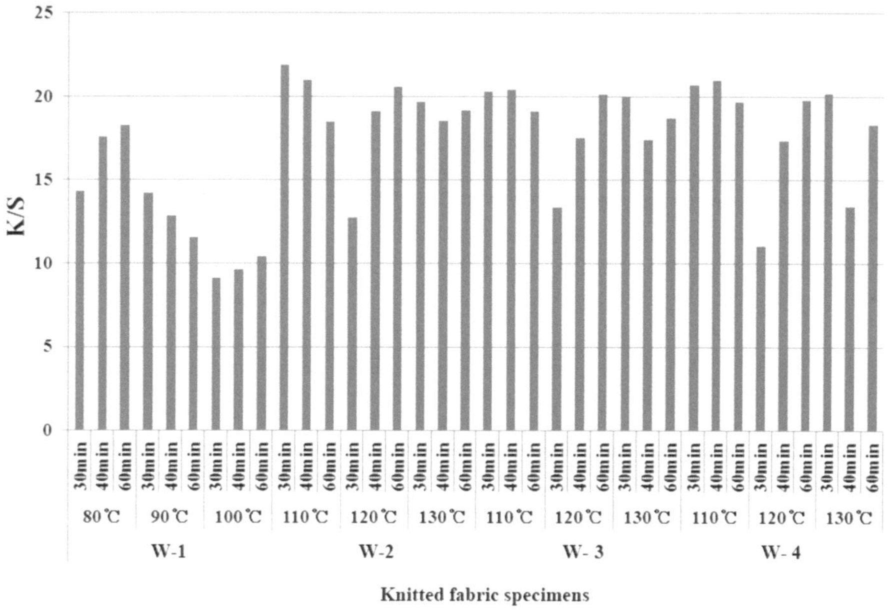


**Figure 5.**
*K/S of the warm-up knitted fabrics according to dyeing condition.*

In the case of nylon/PP ceramic-incorporated fabric (W-1), the K/S decreased with increasing dyeing temperature and increased with increasing dyeing time at dyeing temperatures of 80 and 100°C, but it decreased at a dyeing temperature of 90°C. The ceramic-incorporated and regular PET fabrics (W-2, −3, and − 4) showed a different trend of K/S according to the dyeing temperature and time. The dye affinity of the ceramic-incorporated PET fabrics W-2 and -3 according to the dyeing temperature and time was slightly higher than that of the regular PET fabric (W-4). Furthermore, the dye affinity of the ceramic-incorporated PET fabrics (W-2 and -3) at a dyeing temperature of 110°C decreased with increasing dyeing time. At a dyeing temperature of 120°C, the dye affinity increased with increasing dyeing time. At 130°C, the dye affinity decreased and increased with increasing dyeing time. The optimal dyeing conditions in the dyeing process of the nylon/PP sea and island warm-up knitted fabric were a dyeing temperature of 80°C and a 40 min dyeing time. The optimal conditions for the ceramic-incorporated PET knitted fabrics were 110°C and 30 min, at which K/S showed the highest value.

### *4.3.2 Color fastness to washing of the warm-up knitted fabric specimens*

**Table 5** lists the color fastness to washing of the warm-up knitted fabric specimens according to the dyeing temperature and time. The color fastness to washing of the nylon/PP knitted fabric was superior to that of the ceramic-incorporated and regular PET fabrics, which was superior in the staining fabric, such as acetate and nylon, than cotton, polyester, acrylic, and wool. In particular, the color fastness to washing of the ceramic-incorporated PET-knitted fabrics exhibited a superior grade (4–5) to staining fabrics such as cotton, acrylic, and wool regardless of the dyeing temperature and time.

| Staining fabric | | Acetate | | | | Cotton | | | | Nylon | | | | Polyester | | | | Acrylic | | | | Wool | | | |
|---|---|---|---|---|---|---|---|---|---|---|---|---|---|---|---|---|---|---|---|---|---|---|---|---|---|
| Dyeing temp. (°C) | Dyeing time (min) | W-1 | W-2 | W-3 | W-4 | W-1 | W-2 | W-3 | W-4 | W-1 | W-2 | W-3 | W-4 | W-1 | W-2 | W-3 | W-4 | W-1 | W-2 | W-3 | W-4 | W-1 | W-2 | W-3 | W-4 |
| 80, (110) | 30 | 4–5 | 3 | 3 | 3 | 4–5 | 4–5 | 4–5 | 4 | 4–5 | 2–3 | 2–3 | 2–3 | 4–5 | 3–4 | 3–4 | 4 | 4–5 | 4–5 | 4–5 | 4–5 | 4–5 | 4–5 | 4–5 | 4–5 |
| | 40 | 4 | 3–4 | 3 | 4–5 | 4 | 4–5 | 4–5 | 4–5 | 4 | 3–4 | 2–3 | 4 | 4–5 | 4 | 3–4 | 4–5 | 4–5 | 4–5 | 4–5 | 4–5 | 4–5 | 4–5 | 4–5 | 4–5 |
| | 60 | 4 | 3–4 | 3–4 | 3–4 | 4–5 | 4–5 | 4–5 | 4–5 | 3–4 | 3–4 | 3 | 3 | 3–4 | 4 | 4 | 4 | 4–5 | 4–5 | 4–5 | 4–5 | 4–5 | 4–5 | 4–5 | 4–5 |
| 90, (120) | 30 | 4–5 | 3–4 | 3–4 | 3–4 | 4–4 | 4–5 | 4–5 | 4–5 | 4–5 | 3 | 3 | 3–4 | 4–5 | 4 | 4 | 4 | 4–5 | 4–5 | 4–5 | 4–5 | 4–5 | 4–5 | 4–5 | 4–5 |
| | 40 | 4–5 | 4 | 3–4 | 4–5 | 4–5 | 4–5 | 4–5 | 4–5 | 4–5 | 4 | 3 | 4 | 4–5 | 4–5 | 4 | 4–5 | 4–5 | 4–5 | 4–5 | 4–5 | 4–5 | 4–5 | 4–5 | 4–5 |
| | 60 | 4–5 | 4 | 3–4 | 4–5 | 4–5 | 4–5 | 4–5 | 4–5 | 4–5 | 4 | 3 | 4 | 4–5 | 4–5 | 4 | 4–5 | 4–5 | 4–5 | 4–5 | 4–5 | 4–5 | 4–5 | 4–5 | 4–5 |
| 100, (130) | 30 | 4–5 | 4 | 4 | 4–5 | 4–5 | 4–5 | 4–5 | 4–5 | 4–5 | 4 | 3–4 | 3–4 | 4–5 | 4–5 | 4 | 4–5 | 4–5 | 4–5 | 4–5 | 4–5 | 4–5 | 4–5 | 4–5 | 4–5 |
| | 40 | 4–5 | 4–5 | 3–4 | 4 | 4–5 | 4–5 | 4–5 | 4–5 | 4–5 | 3–4 | 3 | 3–4 | 4–5 | 4–5 | 4 | 4–5 | 4–5 | 4–5 | 4–5 | 4–5 | 4–5 | 4–5 | 4–5 | 4–5 |
| | 60 | 4–5 | 4–5 | 4 | 4–5 | 4–5 | 4–5 | 4–5 | 4–5 | 4–5 | 4 | 3–4 | 4 | 4–5 | 4–5 | 4–5 | 4–5 | 4–5 | 4–5 | 4–5 | 4–5 | 4–5 | 4–5 | 4–5 | 4–5 |

*Note: 80, 90, 100 °C for W-1; 110, 120, 130°C for W-2, 3 and 4.*

**Table 5.**
*Color fastness to washing of the warm-up knitted fabric specimens.*

### 4.4 Physical properties of the cool feeling yarns

According to Huvis Co. Ltd. [30], hydrophilic PET coolness yarn with noncircular and different shaped cross-sections of the constituent filaments in the yarn was spun on the melt spinning machine using a ceramic nanopowder mixed polymer made by adding an intermittently polyol molecule between the molecule chains of TPA and EG. Therefore, the tensile properties of the hydrophilic PET coolness yarn were examined and compared with those of the hydrophobic PET coolness and regular PET yarns. **Figure 6** presents SEM images of cross-sections of the three coolness yarn specimens.

The noncircular and different shaped cross-sections of the constituent filaments in the hydrophilic (C-1) and hydrophobic (C-2) coolness PET yarns were observed. Ceramic nanoparticles in the hydrophilic coolness PET yarn were noted, which were larger than those of the hydrophobic coolness PET yarns. The SEM image of a circular cross-section of each filament in the regular PET yarn (C-3) was slightly distorted because of the false twist in the texturing process. **Table 6** lists the physical properties of the three coolness yarn specimens. The tenacity of the hydrophilic (C-1) and hydrophobic coolness (C-2) PET yarns was lower than that of the regular PET yarn (C-3). This was attributed to the ceramic particles in the yarn. Hence, the boundary characteristics between the base polymer and ceramic particles incorporated in the yarns deteriorated the tenacity of the yarns. The breaking strain of the hydrophilic coolness PET yarn (C-1) was compared with those of the hydrophobic coolness (C-2) and regular PET (C-3) yarns, and was lower than that of C-2 and -3, which was attributed to a similar reason to tenacity.

The wet and dry thermal shrinkages of the yarns need to be considered to examine the process design for the wet and dry heats subjected to the dyeing and finishing processes. The wet and dry thermal shrinkages of the three coolness yarn specimens were listed in **Table 6**. As shown in **Table 6**, the wet and dry thermal shrinkages of

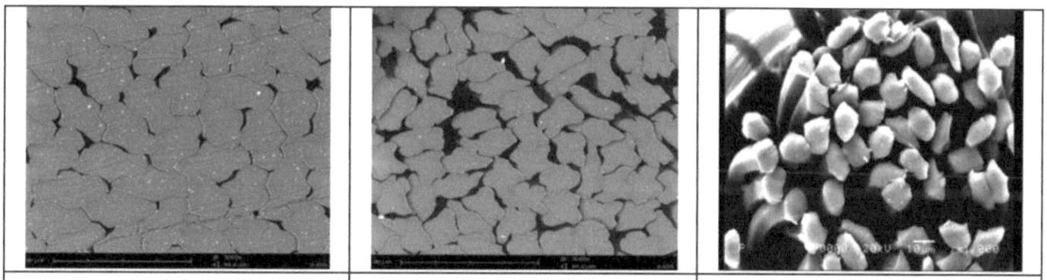

(a) C-1 (b) C-2 (c) C-3

**Figure 6.**
*SEM images of cross-sections of three kinds of coolness yarn specimens. C-1, C-2, (c) C-3.*

| Specimen | Tenacity (g/d) | | Breaking strain (%) | | Thermal shrinkage (%) | | | |
|---|---|---|---|---|---|---|---|---|
| | Mean | Dev. | Mean | Dev. | Mean (D) | Dev. (D) | Mean (W) | Dev. (W) |
| C-1 | 4.182 | 0.758 | 3.371 | 0.741 | 13.280 | 0.623 | 6.674 | 1.186 |
| C-2 | 4.542 | 0.634 | 5.350 | 0.528 | 16.418 | 2.895 | 15.214 | 3.648 |
| C-3 | 4.886 | 0.326 | 5.640 | 0.289 | 9.562 | 1.945 | 4.348 | 0.675 |

*Note: D: dry; W: wet.*

**Table 6.**
*Physical property of coolness yarn specimens.*

the hydrophilic and hydrophobic coolness PET yarns (C-1 and -2) were higher than those of the regular PET (C-3) yarns, respectively. This was attributed to the higher thermal conductivity of the ceramic nanoparticle-incorporated PET yarns than that of the regular PET yarn, which gave these yarns higher thermal stress, resulting in higher thermal shrinkage. In addition, the wet and dry thermal shrinkages of the hydrophilic coolness PET yarn (C-1) were lower than those of the hydrophobic coolness PET yarn (C-2), respectively. This was attributed to the formation of a hydrogen bond between the molecular chain formed by the moisture in the hydrophilic coolness PET yarn, which enabled the hydrophilic yarn to deform less from the internal thermal stress caused by the wet and dry heats, resulting in lower wet and dry thermal shrinkages than the hydrophobic yarn. In addition, which was partly attributed to the larger ceramic particles embedded in the C-1 yarn, as shown in **Figure 6**, than those in the C-2 yarn, which makes less thermal shrinkages.

### 4.5 Contact cool feeling characteristics of the coolness knitted fabrics

Qmax measures the contact cool feeling as human skin contacts the fabric in the transient state. **Figure 7** shows the mean of Qmax of the three coolness knitted fabric specimens according to the dyeing temperature and time.

The Qmax of the hydrophilic (C-1) and hydrophobic coolness (C-2) PET fabrics was higher than that of the regular PET knitted fabric (C-3). This was attributed to the higher thermal conductivity of the ceramic particles in the yarns, which provided higher heat transfer, resulting in a higher Qmax of these knitted fabrics than the regular PET fabric. On the other hand, the Qmax of the hydrophilic coolness PET fabric (C-1) was slightly lower than that of the hydrophobic coolness PET fabric (C-2), which was attributed to hydrogen bonding between the molecular chains in the hydrophilic PET yarn (C-1), and partly attributed to the larger ceramic particles

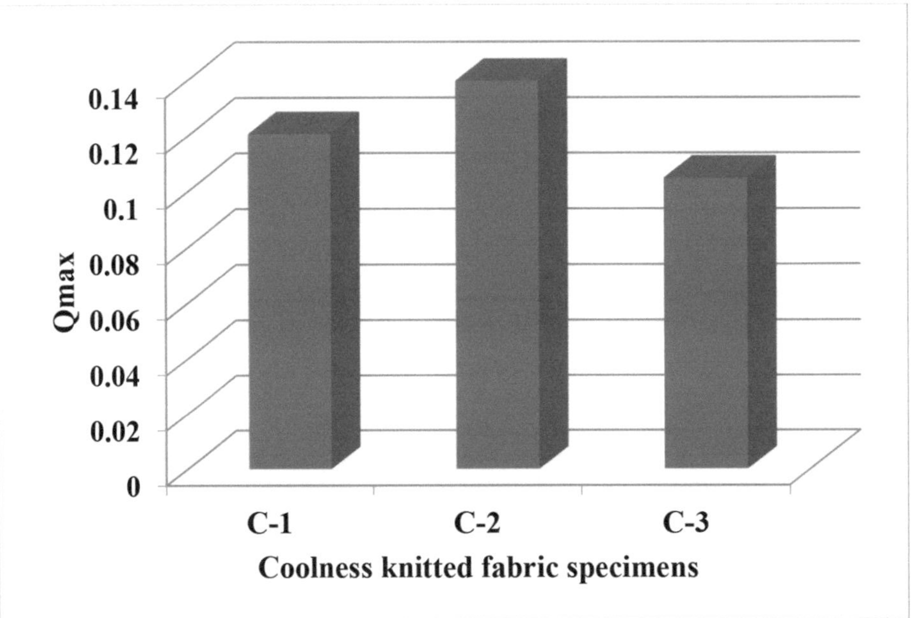


**Figure 7.**
*Mean Qmax of the coolness knitted fabrics.*

embedded in the hydrophilic coolness yarns (C-1) than those in the hydrophobic coolness yarn (C-2). Hence, they prevent heat emanating from the human body from flowing to the outside through the fabric, resulting in a lower Qmax of the hydrophilic coolness PET fabric than the hydrophobic fabric.

**Figure 8** presents the Qmax of the three coolness knitted fabrics according to dyeing temperature and time. A considerable change in the Qmax of the hydrophilic (C-1) and hydrophobic coolness PET (C-2) fabrics according to the dyeing temperature and time was not observed. On the other hand, the Qmax of the three fabric specimens treated with 110°C dyeing temperature (30 and 40 min) was slightly higher than those treated with 120°C and 130°C (30 and 40 min), which means the better coolness feeling of fabrics treated with a 110°C dyeing temperature. Particularly, in the hydrophilic coolness fabric (C-1), the highest Qmax was observed at a dyeing temperature of 110°Cand dyeing time of 30 or 40 min. Hence, a superior coolness feeling was achieved using these dyeing conditions.

## 4.6 Dye-affinity and color fastness characteristics of the coolness knitted fabric specimens

The dye-affinity characteristics of the three knitted fabrics were compared and discussed according to the dyeing temperature and time. **Figure 9** presents the dye-affinity (K/S) of the three knitted fabrics according to the dyeing temperature and time. As shown in **Figure 9**, the K/S of the hydrophilic coolness PET knitted fabric (C-1) was higher than those of the hydrophobic coolness and regular PET knitted fabrics (C-2 and C-3), which was attributed to the higher dissolution speed of the disperse dyestuff due to the hydrophilic property of the C-1 yarn specimen compared to the C-2 and C-3 yarn specimens.

Furthermore, the dye-affinity of the C-1 was accelerated by increasing the diffusion speed into the hydrophilic structure in the yarn. In addition, the K/S of the C-1 showed the highest value at the dyeing temperature, 110°C regardless of the dyeing

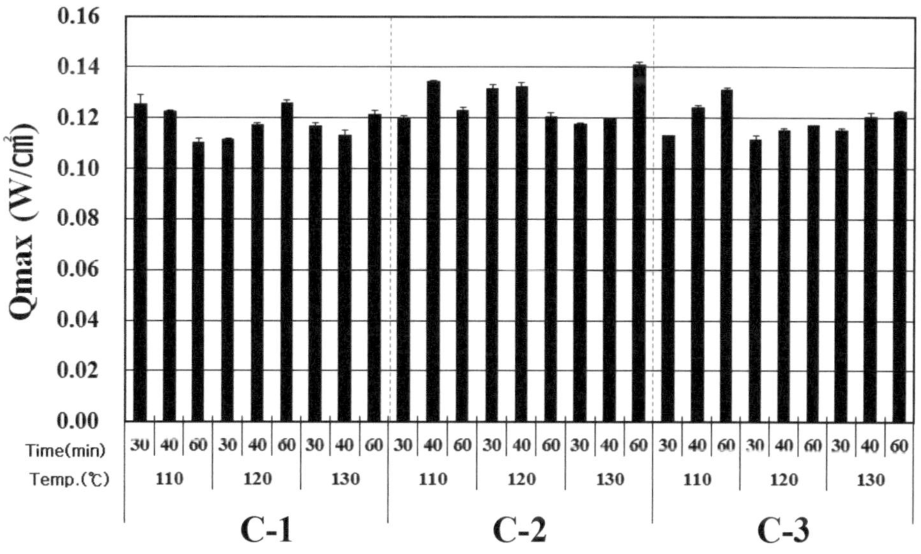


**Figure 8.**
*Qmax of the coolness and regular PET knitted fabrics.*

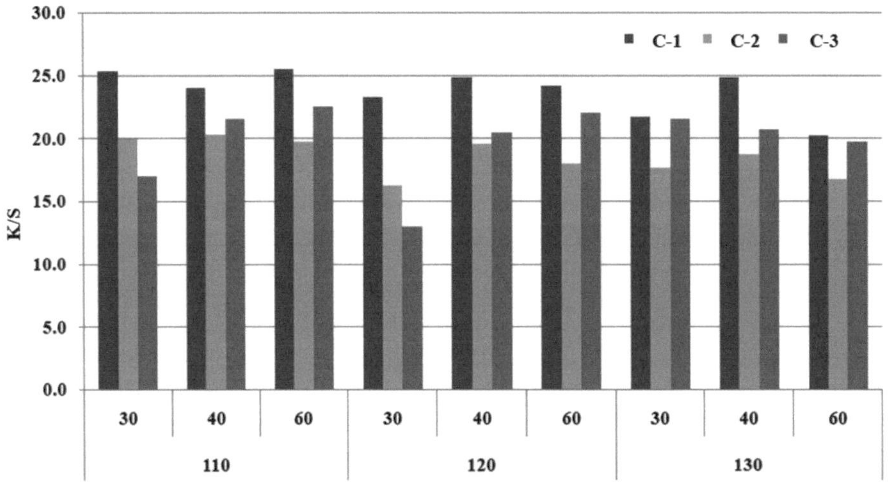


**Figure 9.**
*K/S of the coolness knitted fabric specimens.*

time, and decreased slightly with increasing dyeing temperature, indicating that the hydrophilic coolness PET knitted fabric (C-1) shows high dye-affinity efficiency at the low dyeing temperature. **Figure 10** presents the ΔE of the three knitted fabric specimens according to the dyeing temperature and time. The ΔE of the C-1 was higher than those of the C-2 and C-3 like K/S. This was attributed to a similar reason explained in the K/S. The ΔE of the C-1 and C-2 decreased with increasing dyeing temperature and exhibited the highest values at a dyeing temperature of 110°C, demonstrating high dye-affinity efficiency at the low dyeing temperature of the ceramic-incorporated PET knitted fabrics.

**Table** 7 lists the color fastness to washing of the three knitted fabrics treated with different stain fabrics according to the dyeing temperature and time. The color fastness to washing of the C-1 and C-2 was slightly inferior to C-3 against stain fabrics such as acetate and nylon. The color fastness to washing of the C-1 specimen showed superior values (grade: 4 or 5) and was similar to those of C-2 and C-3 against the stain

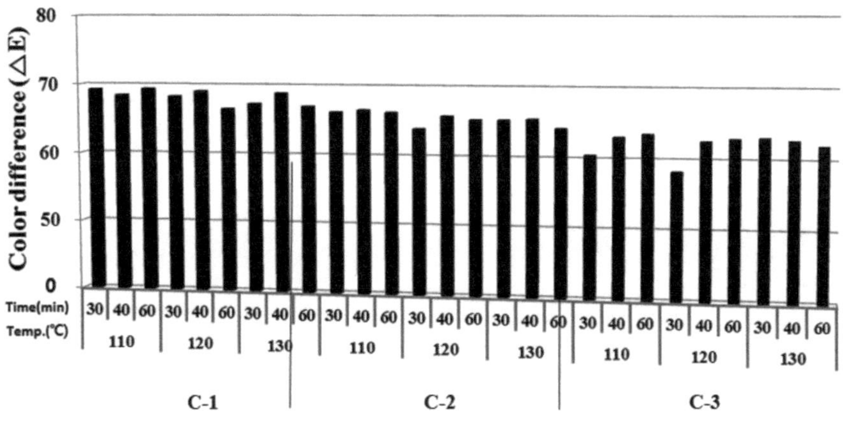


**Figure 10.**
*ΔE of the coolness knitted fabric specimens.*

| | | Acetate | | | Cotton | | | Nylon | | | Polyester | | | Acrylic | | | Wool | | |
|---|---|---|---|---|---|---|---|---|---|---|---|---|---|---|---|---|---|---|---|
| Dyeing temp. (°C) | Dyeing time (min) | C-1 PET | C-2 PET | C-3 PET | C-1 PET | C-2 PET | C-3 PET | C-1 PET | C-2 PET | C-3 PET | C-1 PET | C-2 PET | C-3 PET | C-1 PET | C-2 PET | C-3 PET | C-1 PET | C-2 PET | C-3 PET |
| 110 | 30 | 3–4 | 3–4 | 4–5 | 4–5 | 4–5 | 4–5 | 3 | 3 | 4 | 4 | 4 | 4–5 | 4–5 | 4–5 | 4–5 | 4–5 | 4–5 | 4–5 |
| | 40 | 4–5 | 4–5 | 4–5 | 4–5 | 4–5 | 4–5 | 4 | 4 | 4 | 4–5 | 4–5 | 4–5 | 4–5 | 4–5 | 4–5 | 4–5 | 4–5 | 4–5 |
| | 60 | 4–5 | 4–5 | 4–5 | 4–5 | 4–5 | 4–5 | 3–4 | 4–5 | 3–4 | 4–5 | 4–5 | 4–5 | 4–5 | 4–5 | 4–5 | 4–5 | 4–5 | 4–5 |
| 120 | 30 | 4–5 | 4–5 | 4 | 4–5 | 4–5 | 4–5 | 4 | 3–4 | 3–4 | 4–5 | 4 | 4 | 4–5 | 4–5 | 4–5 | 4–5 | 4–5 | 4–5 |
| | 40 | 4–5 | 4–5 | 4–5 | 4–5 | 4–5 | 4–5 | 4 | 4 | 4–5 | 4–5 | 4–5 | 4–5 | 4–5 | 4–5 | 4–5 | 4–5 | 4–5 | 4–5 |
| | 60 | 4–5 | 4–5 | 4–5 | 4–5 | 4–5 | 4–5 | 4 | 4 | 4–5 | 4–5 | 4–5 | 4–5 | 4–5 | 4–5 | 4–5 | 4–5 | 4–5 | 4–5 |
| 130 | 30 | 4–5 | 4–5 | 4–5 | 4–5 | 4–5 | 4–5 | 4 | 3–4 | 4 | 4–5 | 4 | 4–5 | 4–5 | 4–5 | 4–5 | 4–5 | 4–5 | 4–5 |
| | 40 | 4–5 | 4–5 | 4–5 | 4–5 | 4–5 | 4–5 | 4 | 4 | 4–5 | 4–5 | 4–5 | 4–5 | 4–5 | 4–5 | 4–5 | 4–5 | 4–5 | 4–5 |
| | 60 | 4–5 | 4–5 | 4–5 | 4–5 | 4–5 | 4–5 | 4 | 4 | 4–5 | 4–5 | 4–5 | 4–5 | 4–5 | 4–5 | 4–5 | 4–5 | 4–5 | 4–5 |

**Table 7.**
*Color fastness to washing of the coolness fabric specimens.*

fabrics, such as cotton, polyester, acrylic, and wool, regardless of the dyeing temperature and time.

## 5. Summary

The thermal wear comfort of the warm-up knitted fabric made from the ceramic-incorporated nylon/PP sea and island filament was examined, and the dye-affinity of the nylon/PP-knitted fabric was also investigated according to the dyeing temperature and time. The tenacity and breaking strain of the nylon/PP composite yarn were approximately 4gf/d and 4.5%, respectively, meaning that there was no problem in weavability in the weaving process. The dry and wet thermal shrinkages of the nylon/PP composite yarns ranged from 18 to 19%, which were higher than those of the regular PET yarns. The Qmax of the nylon/PP warm-up knitted fabric was lower than that of the regular PET knitted fabric, which was attributed to the greater heat emitted from the ceramic particles incorporated in the core of the nylon/PP composite yarns. In addition, the TIV of the nylon/PP knitted fabric was approximately 9.3% higher than that of the regular PET knitted fabric, which was also attributed to the greater heat emanating from the ceramic particles incorporated in the island of the nylon/PP sea and island yarn than PET yarn. Considering energy savings in the dyeing process, the optimal dyeing condition from the K/S dye-affinity result of the nylon/PP ceramic-incorporated knitted fabric was a dyeing temperature of 80°C and a dyeing time of 40 min. For the ceramic-incorporated PET knitted fabrics, the optimal dyeing condition was 110°C and 30 min. The color fastness to washing of the nylon/PP knitted fabric was superior to that of the ceramic-incorporated and regular PET fabrics for all types of staining fabrics, which was superior for staining fabrics, such as acetate and nylon, than cotton, PET, acrylic, and wool.

On the other hand, the contact coolness feeling (Qmax) and dyeability characteristics of the ceramic incorporated and hydrophilic coolness feeling PET knitted fabric were examined in terms of the dyeing temperature and time. The tenacity and breaking strain of the ceramic incorporated hydrophilic coolness PET yarn were approximately 4.18 g/d and 3.4%, respectively, which was estimated as being no problem in weavability in the weaving process. The dry and wet thermal shrinkages of the hydrophilic coolness PET yarn were 13.3 and 6.7%, respectively, which was 1.3 to 1.5 times higher than those of the regular PET yarn. The Qmax of the ceramic incorporated hydrophilic and hydrophobic coolness PET knitted fabrics were higher than that of the regular PET fabric, indicating the superior contact coolness feeling of these fabrics. The maximum value of Qmax of the hydrophilic coolness PET knitted fabric according to the dyeing temperature and time appeared at a dyeing temperature of 110°C and dyeing time of 30 or 40 min, that is, exhibited superior contact coolness feeling at this dyeing condition. Considering the energy saving and dyeability of the newly developed yarn and fabric in the dyeing process, the K/S of the ceramic-embedded hydrophilic coolness PET knitted fabric decreased with increasing dyeing temperature, irrespective of the dyeing time, and higher than those of the hydrophobic coolness and regular PET fabrics. In addition, the maximum K/S value of the ceramic-embedded hydrophilic coolness PET fabric according to the dyeing temperature and time was observed at 110°C and 30 min or 40 min, indicating the optimal dyeing conditions for high dye-affinity and low energy consumption. The $\Delta E$ of these fabric specimens according to dyeing temperature and time showed a similar trend to the K/S. The color fastness to washing of the ceramic-embedded hydrophilic coolness knitted fabric was slightly inferior to the regular PET fabric in the case of stain fabrics such as acetate and nylon, whereas it was superior in the case of stain fabrics, such as cotton, polyester, acrylic, and wool, regardless of the dyeing temperature and time.

## 6. Concluding remarks

Application areas of ceramic particles to textiles are known as water repellence, anti-bacteria, flame retardation, anti-static, and UV protection. Concerning water repellence, nano-tex improved the water-repellent property by creating nano-whiskers of hydrocarbons, which are added to the fabric to create a peach fuzz effect [58]. The performance is permanent while maintaining breathability, and a hydrophobic property can be imparted on the fabrics by coating it with a thin nanoparticulate plasma film [70, 71]. For imparting anti-bacterial properties, titanium dioxide ($TiO_2$) [72–74] and zinc oxide [53] are used. Fabrics treated with nano-$TiO_2$ could provide effective protection against bacteria and the discoloration of stains. Nano-ZnO provides effective photocatalytic properties like $TiO_2$, once it is illuminated by light, imparting anti-bacterial properties to textiles [75–77]. Flame-retardant (FR) properties of textiles have been explored using special flame-retardant fibers such as modacrylic (Kanekalon, SNIA), fire-retardant viscose rayon (FR-rayon), and anti-static PET fiber, Beltron, has been used in protective clothing for FR and anti-static textile materials [78–80]. On the other hand, many attempts to obtain flame retardant activity by the formation of a layer of ceramics (glass, $SiO_2$, $Al_2O_3$) have been explored [81–84]. As mentioned previously, abundant research works [56, 57, 61–66] concerning the improvement of the anti-static properties of textiles were conducted using ceramic particles such as ZnO [62, 65, 66] and ATO [56, 57, 61, 63–66]. In addition, ceramic UV blockers are usually certain semiconductor oxides such as $TiO_2$, ZnO, $SiO_2$, and $Al_2O_3$. Of these, $TiO_2$ [52, 63, 64, 66] and ZnO [53–55, 65, 66] are commonly used as previously explained. In particular, nano-sized $TiO_2$ and ZnO are more efficient at absorbing and scattering UV radiation, and several studies on their application of UV protection to fabric were conducted [51–55, 63–66]. The application of various ceramic materials to textiles has been highlighted in many papers, and new technology to overcome the limitation of conventional methods will be of practical use for engineering various workwear protective clothings, including fashioned textile materials.

## Author details

Seung Jin Kim[1*] and Hyun Ah Kim[2]

1 Department of Fiber System Engineering, Yeungnam University, Kyeongsan, South Korea

2 Korea Research Institute for Fashion Industry, Daegu, South Korea

*Address all correspondence to: sjkim@ynu.ac.kr

## References

[1] Unitika's activity fields: connected through technologies and principles [Internet]. 2022. Available from: https://www.unitika.co.jp/e/company/img-pdf/download/p03-04.pdf

[2] Kurray. Fabric, manmade leather, nonwoven fabric, hook and loop fasteners [Internet]. 2022. Available from: http://www.kuraray.com/products/fiber

[3] Mitsubishi Rayon Carbon Fiber and Composites, Inc. Safety data sheet [Internet]. 2022. Available from: http://mrcfac.com/wp-content/uploads/2013/09/SDS-Grafil-precision_cut-unsize-_05-17-12.pdf?1475465251

[4] Toray [Internet]. 2022. Available from: https://www.toray.com/global/technology/download

[5] Seiren KB. Flethermo [Internet]. 2022. Available from: https://www.kbseiren.com/english/pro-flethermo.html

[6] Lee BS, Nam SI, Song JW, Rho YH. Development trend of functional textile materials technology of coolness/warm up. Fiber Technology and Industry. 2012;**16**:77-84

[7] Rho YH, Kwon OK. Heat keeping warm-up textiles. Fiber Technology and Industry. 2010;**14**:245-255

[8] Jin BY, Yoon HS, Kim DS. Assessment of warm-up textiles. Fiber Technology and Industry. 2012;**16**:89-98

[9] Roh SY. Patent Status of Development of Warm & Cool Biz Functional Textile Goods. Daegeon: The Office of Patent Administration; 2008

[10] Kim HA, Kim SJ. Hand and mechanical properties of nylon high hollow fabrics in relation to the elution characteristics. Journal of the Textiles Institute. 2018;**2**:160-171. DOI: 10.1080/00405000.2017.1334751

[11] Kim HA, Kim SJ. Moisture and thermal permeability of the hollow textured PET imbedded woven fabrics for high emotional garments. Fibers and Polymers. 2016;**17**:427-438. DOI: 10.1007/s12221-016-5942-9

[12] Materials of Warm BIZ and Cool BIZ [Internet]. 2022. Available from: http://www.textopia.or.kr

[13] Development trend of wicking perspiration & quick dry functional textiles [Internet]. 2022. Available from: http://www.textopia.or.kr

[14] Rho HK. Wicking Perspiration and Quick Dry Fabrics [Thesis]. Korea: Information note: Kofoti; 2006

[15] Negishi N, Kikuchi M. Infrared ray effects in biological systems. Ceramic Japan. 1988;**23**:335-339

[16] Furuta T, Shimizu Y, Kondo Y. Evaluating the temperature and humidity characteristics of solar energy absorbing and retaining fabric. Textile Research Journal. 1996;**66**:123-130. DOI: 10.1177/004051759606600З

[17] Bahng GW, Lee JD. Development of heat-generating polyester fiber harnessing catalytic ceramic powder combined with heat-generating super microorganisms. Textile Research Journal. 2014;**84**:1220-1230. DOI: 10.1177/0040517513501350

[18] Anderson DM, Fessler JR, Pooley MA, Seidel S, Hamblin MR, Beckhan HW, et al. Infrared radiative properties and thermal modeling of

ceramic-imbedded textile fabric. Biomedical Optics Express. 2017;**8**: 1698-1711. DOI: 10.1364/BOE.8.001698

[19] Pooley MA, Anderson DM, Beckham HW, Brennan JF. Engineered emissivity of textile fabrics by the inclusion of ceramic particles. Biomedical Optics Express. 2016;**24**: 10556-10564. DOI: 10.1364/OE.24.010556

[20] Kim HA, Kim SJ. Heat storage and release characteristics of ceramic-imbedded woven fabric for emotional clothing. Autex Research Journal. 2019; **19**:165-172. DOI: 10.1515/aut-2018-0039

[21] Kim HA, Kim SJ. Wear comfort properties of $ZrC/Al_2O_3$/graphite embedded heat storage woven fabrics for garments. Textile Research Journal. 2019;**89**:1394-1407. DOI: 10.1177/004051751877068

[22] Matsudaira M, Kondo Y. The effect of a grooved hollow in a fibre on fabric moisture and heat-transport properties. Journal of the Textile Institute. 1996;**87**: 409-416. DOI: 10.1080/00405009608631342

[23] Onofrei E, Rocha AM, Catarino A. The influence of knitted fabrics' structure on the thermal and moisture management properties. Journal of Engineered Fibers and Fabrics. 2011;**6**: 10-22. DOI: 10.117/155892501100600403

[24] Supuren G, Oglakcioglu N, Ozdil N, Marmarali A. Moisture management and thermal absorptivity properties of double-face knitted fabrics. Textile Research Journal. 2011;**81**:1320-1331. DOI: 10.1177/0040517511402122

[25] Majumdar A, Mukhopadhyay S, Yadav R. Thermal properties of knitted fabrics made from cotton and regenerated bamboo cellulosic fibres. International Journal of Thermal Sciences. 2010;**49**:2042-2048. DOI: 10.1016/j.ijthermalsci.2010.05.017

[26] Pac MJ, Bueno M, Renner M, Kasmi SE. Warm-cool feeling relative to tribological properties of fabrics. Textile Research Journal. 2001;**71**:806-812. DOI: 10.1177/004051750107100910

[27] Stankovic SB, Popovic D, Poparic GB. Thermal properties of textile fabrics made of natural and regenerated cellulose fibers. Polymer Testing. 2008; **27**:41-49. DOI: 10.1016/j.polymertesting.2007.08.003

[28] Choi K, Cho G, Kim P, Cho C. Thermal storage/release and mechanical properties of phase change materials on polyester fabrics. Textile Research Journal. 2004;**74**:292-302. DOI: 10.1177/004051750407400402

[29] Shim H, McCullough EA, Jones BW. Using phase change materials in clothing. Textile Research Journal. 2001; **71**:495-504. DOI: 10.1177/004051750107100605

[30] Choi K, Chung H, Lee B, Chung K, Cho G, Park M, et al. Clothing temperature changes of phase change material-treated warm-up in cold and warm environments. Fibers and Polymers. 2005;**6**:343-347. DOI: 10.1007/BD02875673

[31] Mondal S. Phase change materials for smart textiles – An overview. Applied Thermal Engineering. 2008;**28**: 1550-1558. DOI: 10.1016/j.applthermaleng.2007.08.009

[32] Kim HA, Kim SJ. Moisture responded transformable property of intelligent coolness knitted fabrics for sportswear clothing. Autex Research Journal. 2017;**17**:250-258. DOI: 10.1515/aut-2016-0013

[33] Mather RR. Intelligent Textiles. Re view of Progress in Coloration and Rela ted Topics. 2001;**31**:36-41

[34] Mikucioniene D, Bivainyte A. Investigation on dynamic of heat transfer through double–layered weft knits. Autex World Textile Conference. 2013;**13**:14

[35] Piraee M, Saharkhiz S. The influence of structural variations of weft knitted fabrics on capacity of heat transfer of fabrics. Autex World Textile Conference. 2013;**13**:85

[36] Yanılmaz M, Kalaoğlu K. Investigation of wicking, wetting and drying properties of acrylic knitted fabrics. Textile Research Journal. 2012; **82**:820-827

[37] Troynikov O, Wardiningsih W. Moisture management properties of wool/polyester and wool/bamboo knitted fabrics for the sportswear base layer. Textile Research Journal. 2011;**81**: 621-630. DOI: 10.1177/0040517510392461

[38] Chen Q, Fan J, Sarkar MK. Biomimetics of branching structure in warp knitted fabrics to improve water transport properties for comfort. Textile Research Journal. 2012;**82**:1131-1140. DOI: 10.1177/0040517512438

[39] Chen Q, Fan J, Sarkar MK, Jiang G. Biomimetics of plant structure in knitted fabrics to improve the liquid water transport properties. Textile Research Journal. 2010;**80**:568-568. DOI: 10.1177/0040517509340600

[40] Chen Q, Fan J, Sarkar MK, Bal K. Plant-based biomimetic branching structures in knitted fabrics for improved comfort-related properties. Textile Research Journal. 2011;**81**:1039-1049. DOI: 10.1177/004051751039757

[41] Fan J, Sarkar M, Szeto Y, Tao X. Plant structured textile fabrics. Materials Letters. 2007;**61**:561-570. DOI: 10.1016/j.matlet.2006.05.009

[42] Sarkar M, Fan J, Szeto Y, Tao X. Biomimetics of plant structure in textile fabrics for the improvement of water transport properties. Textile Research Journal. 2009;**79**:657-668. DOI: 10.1177/0040517508095604

[43] Sarkar M, Fan J, Szeto Y, Tao X. Development and characterization of light weight plant structured fabrics. Fibers and Polymers. 2009;**10**:343-350. DOI: 10.1007/s12221-009-0343-y

[44] Kim HA. Water/moisture vapor permeabilities and thermal wear comfort of the Coolmax ® /bamboo/tencel included PET and PP composite yarns and their fabrics. Journal of Textile Institute. 2021;**112**:1940-1953. DOI: 10.1080/00405000.2020.1853409

[45] Kim HA. Wear comfort of woven fabrics for clothing made from composite yarns. Fibers and Polymers. 2021;**22**:2344-2353. DOI: 10.1007/s12221-021-0562-4

[46] Kim HA. Moisture vapor permeability and thermal wear comfort of ecofriendly fiber-embedded woven fabrics for high-performance clothing. Materials. 2021;**14**: 6205-6228. DOI: 10.3390/ma14206205

[47] Kim HA. Eco-friendly fibers embedded yarn structure in high-performance fabrics to improve moisture absorption and drying properties. Polymers. 2023;**15**:581-602. DOI: 10.3390/polym15030581

[48] Kim HA. Wear comfort of heat storage/release fabrics containing $Al_2O_3$/graphite yarns. Fibers and Polymers. 2021;**23**:554-564. DOI: 10.1007/s12221-021-0312-7

[49] Kim HA. Heat release and wear comfort characteristics of the ceramic imbedded fabrics for cold weather protective clothing. Journal of Industrial Textiles. 2022;**52**:1-21. DOI: 10.1177/15280837221109638

[50] Kim HA. Wear comfort and thermal insulation of ceramic-imbedded fabrics with different yarn structures by thermal manikin experiment. International Journal of Clothing Science. 2022;**35**:197-213. DOI: 10.1108/IJCST-06-2021-0081

[51] Xin JJ, Daoud WA, Kong YY. A new approach to UV-blocking treatment for cotton fabrics. Textile Research Journal. 2004;**74**:97-100. DOI: 10.1177/004051750407400202

[52] Yang HY, Zhu SK, Pan N. Studying the mechanisms of titanium dioxide as ultraviolet-blocking additive for films and fabrics by and improved scheme. Journal of Applied Polymer Science. 2003;**92**: 3201-3210. DOI: 10.1002/app.20327

[53] Saito M. Antibacterial, deodorizing and UV absorbing materials obtained with zinc oxide (ZnO) coated fabrics. Journal of Coated Fabrics. 1993;**23**: 150-164. DOI: 10.1177/152808379302300205

[54] Xiong MN, Gu GX, You B, Wu LM. Preparation and characterization of poly (styrenebutylacrylate) latex/nano-ZnO nanocomposites. Journal of Applied Polymer Science. 2003;**90**:1923-1931. DOI: 10.1002/app.12869

[55] Wang RH, Xin JJ, Tao XM, Daoud WA. ZnO nanorods grown on cotton fabrics at low temperature. Chemical Physics Letters. 2004;**398**: 250-255. DOI: 10.1016/j.cplett.2004.09.077

[56] Dang HP, Luc QH, Le T, Le VH. The optimum fabrication condition of p-type antimony tin oxide thin films prepared by DC magnetron sputtering. Journal of Nanomaterials. 2016;**687**:1012-1022. DOI: 10.1155/2016/7825456

[57] Wu Y, Chi YB, Nie JX, Yu AP, Chen XH. Preparation and application of novel fabric finishing agent containing nano ATO. Journal of Functional Polymers. 2002;**15**:43-47. DOI: 10.5539/ijc.v1n1p18

[58] Wong YWH, Yuen CWN, Leung MYS, Ku SKA, LAM HLI. Selected applications of nano-technology in textile. Autex Research Journal. 2006;**6**:1-8

[59] Li Y, Wu DX, Hu JY, Wang SX. Novel infrared radiation properties of cotton fabric coated with nano Zn/ZnO particles. Colloids and Surfaces. 2007; **300**:140-144. DOI: 10.1016/j.colsurfa.2007.01.001

[60] Ahn TX, Tung DT, Nhan DQ, Hoang TV, Trung DQ, Thu LD, et al. Study of ATO nanoparticles by the solvothermal method for thermal insulated coated glass: A green energy application. Green Process Synthesis. 2016;**5**:529-535. DOI: 10.1515/gps-2016-0068

[61] Sun H, Liu X, Liu B, Yin Z. Preparation and properties of antimony doped tin oxide nanopowders and their conductivity. Material Research Bullitin. 2016;**83**:354-359. DOI: 10.1016/j.materresbull.2016.06.028

[62] Zhou ZW, Chu LS, Tang WM, Gu LX. Studies on the antistatic mechanism of tetrapod-shaped zinc oxide whisker. Journal of Electrostatics. 2003;**57**:347-354. DOI: 10.1016/S0304-3886(02)00171-7

[63] Kim HA. Ultra-violet protection and anti-static characteristics with heat release/shiedling of $Al_2O_3$/ATO/$TiO_2$-

imbedded high performance fabrics. Materials. 2022;**15**:3652-3665. DOI: 10.3390/ma15103652

[64] Kim HA. Wear comfort characteristics of $Al_2O_3$/ATO/$TiO_2$-embedded multi-functional PET fabrics. Materials. 2022;**15**:8799-8815. DOI: 10.3390/ma15248799

[65] Kim HA. Investigation of moisture vapor permeability and thermal comfort properties of ceramic embedded fabrics for protective clothing. Journal of the Industrial Textiles. 2022;**52**:1-27. DOI: 10.1177/15280837221142231

[66] Kim HA. Multifunctional characteristics of various inorganic particles imbedded fabrics for workwear protective clothing in cold weather region. Materials

[67] Fan J, Hunter L. Engineering Apparel Fabrics and Garment. Cambridge: The Textile Institute. Woodhead Publishing Limited, CRC press; 1996. pp. 251-260

[68] Xin JH, Formulation CC, Xin JH. Total Color Management in Textiles. Cambridge: Woodhead Publishing Limited; 2006. pp. 136-159

[69] Bona M. Textile Quality, –Physical Methods of Product and Process Control. Biella, Italy: Texilia; 1994

[70] Kim HA. Moisture vapor resistance of coated and laminated breathable fabrics using evaporative wet heat transfer method. Coatings. 2021;**11**: 1157-1265. DOI: 10.3390/coatings11101157

[71] Kim HA. Water repellency/proof/vapor permeability characteristics of coated and laminated breathable fabrics for outdoor clothing. Coatings. 2022;**12**: 12-21. DOI: 10.3390/coatings12010012

[72] Burniston N, Bygott C, Stratton J. Nano technology meets titanium dioxide. Surface Coatings International Part A: Coatings Journal. 2004;**87**: 179-814

[73] Sherman, Jonathan, Nanoparticlulate Titanium Dioxide Coatings, and Processes for the Production and Use Thereof, US patent. No. 736738, 2003

[74] Daoud WA, Xin JH. Nucleation and growth of anatase crystallites on cotton fabrics at low temperatures. Journal of the American Ceramic Society. 2004;**87**: 953-955. DOI: 10.1111/j.1551-2916.2004.00953.x

[75] Daoud WA, Xin JH. Low temperature sol-gel processed photocatalytic titania coating. Journal of Sol-Gel Science and Technology. 2004; **29**:25-29. DOI: 10.1023/B:JSST.0000016134.19752.b4

[76] Cui SY, Zu YD, Hui HQ, Zhang JY. Study on anti-bacteria properties of nano-ceramics. Journal of Hebei University of Science and Technology. 2003;**24**:19-22. DOI: 10.3390/nano13030583

[77] Wang RH, Xin JH, Yang Y, Liu HF, Xu LM, Hu JH. The characteristics and photocatalytic activities of silver doped ZnO nanocrystallites. Applied Surface Science. 2004;**227**:312-317. DOI: 10.1016/j.apsusc.2003.12.012

[78] Kim HA, Kim SJ. Flame-retardant and wear comfort properties of modacrylic/FR-rayon/anti-static PET blend yarns and their woven fabrics for clothing. Fibers and Polymers. 2018;**19**: 1869-1879. DOI: 10.1007/s12221-018-1087-3

[79] Kim HA, Kim SJ. Flame retardant, anti-static and wear comfort properties of modacrylic/excel®/anti-static PET

blend yarns and their knitted fabrics. Journal of the Textile Institute. 2019;**110**: 1318-1328. DOI: 10.1080/ 00405000.2019.1565626

[80] Kim HA. Tactile hand and wear comfort of flame-retardant rayon/anti-static polyethylene terephthalate imbedded woven fabrics. Textile Research Journal. 2019;**89**:4658-4669. DOI: 10.1177/0040517519837729

[81] Horrocks AR. Textiles. In: Horrocks AR, Prince D, editors. Fire Retardant Materials. Cambridge: Woodhead Publishing; 2001. pp. 121-181

[82] Perepelkin KE. Chemical fibers with specific properties for industrial application and personnel protection. Journal of Industrial Textiles. 2001;**31**: 87-102. DOI: 10.1106/XU8H-C5J5-8BLT-2EAO

[83] Tolbert TW, Dugan JS, Jaco P, Hendrix JE. Springs Industries, Fire Barrier Fabrics. Vol. 4. US Patent Office; 1989. p. 333174

[84] Hribernik S, Smole MS, Kleinschek KS, Bele M, Jamnik J, Gaberscek M. Flame retardant activity of SiO2-coated regenerated cellulose fibres. Polymer Degradation and Stability. 2007;**92**:1957-1965. DOI: 10.1016/j. carbpol.2017.08.111